D1575670

WILDCATTER

WILDCATTER

the Story of
Michel T. Halbouty
and the Search for Oil

JACK DONAHUE

McGRAW-HILL BOOK COMPANY
New York • St. Louis • San Francisco
Düsseldorf • London • Mexico • Sydney • Toronto

B
Halbouty

2 3 4 5 6 7 8 9 *FGRFGR* 7 8 3 2 1 0 9

LIBRARY OF CONGRESS CATALOGING IN PUBLICATION DATA
Donahue, Jackson.
    Wildcatter.
    Includes index.
    1. Halbouty, Michel Thomas, 1909-   2. Petroleum engineers—
United States—Biography.   I. Title.
TN140.H24D66   622'.33'80924 [B]   79-561
ISBN 0-07-017542-X

Book design by Marsha Picker.

"Wildcattin'? All it takes is guts and acreage. It seems to help some, too, if you're smart and lucky. . . ."

—COLUMBUS MARION (DAD) JOINER

one

MICHEL THOMAS HALBOUTY strode off the Texas A&M University campus on graduation day, 1931, and six weeks later "discovered" an oilfield that has yielded 135 million barrels of petroleum.

The bonanza was near the village of High Island on the Bolivar Peninsula, a slender finger thrust into the Gulf of Mexico from the Texas Coast. It was a steaming marshland that summer, hot and humid with a decisive sun that seemed to rise at full strength and remain so until it set without lingering on the horizon. Seagulls and shore birds were languid. Only men and mosquitoes were energetic, the men because they feared loss of their jobs in that harsh depression year. The mosquitoes swept around them in black swarms, thirsting for blood and carrying debilitating fevers.

Halbouty arrived as a chain-puller on a surveying crew, the meanest, lowest-paid job in the oilfields of that day. It was the only oilfield job he could find though he held Bachelor and Master of Science degrees in both geology and petroleum engineering. He was grateful for *any* oilfield job; his ambition was as obvious as his vanity and bellicosity, and it assured him that a bright future was just around the next clump of bushes.

Slogging through the mud, he pulled his measuring chain behind a burly crewman who cleared a path through the damp brush with a flashing machete. When the measuring chain reached its full length of one hundred feet, Halbouty looked back for signals from the surveyor who was guiding him by sighting through his surveying instrument. He moved about until the surveyor signaled that he was standing on the correct spot. There he drove a stake in the ground and resumed his march, a lean, broad-shouldered powerful young man of 21 who stood just a shade under six feet in khaki shirt and trousers caked white with salt from his perspiration. His black eyes and perpetual tan were inheritances from his Lebanese ancestors, and his oiled black hair, parted fashionably in the middle, gleamed like patent leather. He wore a moustache in the belief it made him appear more mature.

Earlier in the year his employer, the Yount-Lee Oil Company of nearby Beaumont, Texas, had acquired approximately twenty-two hundred acres of leases near High Island village, and it was the surveying crew's job to verify the property's metes and bounds. High Island was not an island at all. It was so named because its elevation of forty-seven feet made it the highest point between Point Bolivar, at the tip of the peninsula, and Sabine Pass, the peninsula's inland extremity.

High Island's elevation was caused by a giant shaft of salt perhaps thirty thousand feet long and more than a mile in diameter. It had been born in the saline residue of an ancient sea. Layer after layer of sediment had been deposited on the salt bed through the ages. Weight of the sediment exerted great pressure on the salt bed. The pressure and extreme heat at that depth made it plastic. It sought release through the weakest spots in the formations above it, moving upward through eons until it halted some thirteen hundred feet below the earth's surface. The stalk of salt had fractured and tilted formations in its rise; it had "uplifted" those near the surface it had not penetrated.

It was the salt dome, as such geological phenomena were called, that had brought Yount-Lee to High Island. In 1901 the country's first great gusher had been drilled on the crest of a salt dome called Spindletop near Beaumont. Spindletop mound was about a mile in diameter and rose some fifteen feet above the general level of the prairie. Inflammable gases escaped from its soil, and water from its several springs was strong with the taste of sulphur.

Because Spindletop became a great oilfield, oilmen sought and found dozens of salt domes along the Texas-Louisiana Gulf Coast in the years that followed. Some salt dome areas became prolific producing fields, some were marginal, and others were barren, but their lure did not diminish.

After Spindletop, petroleum geologists learned that oil migrated from its source beds through porous strata until it reached what they called structural traps, points at which it could go no farther. Generally it migrated upward, riding atop salt water from prehistoric oceans, seeking (like the salt) a release from the pressure of formations above it. The oil accumulated in sand strata or porous rock against the structural traps and an oil "pool" was formed.

Besides salt domes, the most common structural traps were anticlines and faults. An anticline is an arch with the strata dipping on either side, as from the ridge of a house. A fault is a break or dislocation of strata, an interruption in their continuity so that on either side they are elevated or depressed or horizontally disengaged.

A salt dome, then, was a sort of anticline, cherished because it could trap oil around all of its flanks and above the salt itself—if oil there was to be trapped. So High Island had not escaped the oil seekers. Between 1901 and Halbouty's arrival in 1931, ninety-eight wells had been drilled on the crest and flanks of the dome. Only a few had been productive and no more than five hundred thousand barrels of oil had been extracted.

Even Yount-Lee had drilled six dry holes at High Island in 1923, but the company's leader, Miles Frank Yount, had not lost faith in the area. His skill, daring, and unfailing optimism had made Yount-Lee one of the most successful independent oil companies in the world. In 1925 he had created a second great boom at Spindletop by finding new and even more prolific production on the old dome. He had drilled on leases obtained from a brilliant oilman, Marrs McLean.

He also had obtained most of his High Island leases from McLean, and now, in June 1931, High Island was alive again. Several wells were being drilled and preparations for others were underway. As Halbouty pulled his measuring chain through the marsh he watched the rig-builders fashion derricks of heavy timbers, clambering through the latticework they were creating to drive thick spikes into derrick legs and crossbars. Because the ground off the crest of the dome was so soft, a derrick's legs and floor rested on timber "mats" that would not sink in the mud. Catwalks led from the derricks to higher, dryer ground.

Halbouty was no stranger to an oilfield. His hometown was Beaumont, and as a teenager he had worked as a waterboy and truck swamper in the fabled Spindletop field during the second boom—Yount's boom. So wild had been the drilling pace that derricks stood side by side. Halbouty could traverse almost the entire field without putting foot to ground as he carried his water bucket from crew to crew.

Still he watched the action around him as if he had never seen anything like it before. His job was wearying but undemanding; he could freely observe the drilling equipment being placed near and on the derrick floor—the boilers that fired the engine that would power the rotary table that would spin the drill pipe into the earth; the draw-works that would lift the heavy string of pipe to the surface when necessary and be used in other operations; the various pumps and hoses and catlines.

But he kept reminding himself that he was a geologist—a brilliant one, his professors at Texas A&M had said. He felt a special kinship for the earth. Plodding along he smelled the exuding sulphurous water. He dug into the earth and found paraffin mixed in the sand and clay, another indicator of potential oil territory. He was able to delineate where the crest of the dome became the flanks: the vegetation on the crest was of different size, texture, and color than that on the flanks. As he studied the earth's surface he sometimes felt that he was on the verge of seeing *below* the surface. He was like an intelligent dog aching for the power of speech to express his love for his master.

At night the derrick lights and gas flares from older producing wells lit up the sky like a giant carnival. And the noise never ceased. The clanging of metal on metal, the roar of machinery as bits were driven deep into the earth, lured Halbouty to the rigs even when he was bone tired. He would stand in the steaming night and watch the driller and his crew working together with the precision and coordination of a ballet team. And over it all, seeming to permeate even the wood and metal, was the rotting, swampy odor each man inhaled as he worked.

Nothing could pull Halbouty away from a rig when the crew was "pulling a core"—taking a sample from the bore hole. Normally a driller pulled a core only when it was thought he had drilled to a formation that might contain traces of oil. He would send a hollow tool with a specially designed cutting bit to the bottom of the hole. As the bit rotated in the formation, a sample of the stratum would pack into the tool; a "seat" on the tool would close as the assembly was removed from the hole, and the contents of the assembly could be examined when it was opened on the rig floor.

Halbouty would wait and watch with an intensity that made his stomach churn. He had never seen a core containing oil sand as it was taken from the hole, and he wanted desperately to do so. On every occasion he was disappointed.

Electronic devices that could be lowered into a bore hole and could make an accurate "reading" of every formation through which the bit had passed were not yet in use on the Gulf Coast in 1931. Instead, the driller kept a "log" of the bit's passage. He would note in his log, for example, that between 1,000 feet and 1,020 feet the bit had encountered shale, that from 1,020 feet to 1,035 feet it had encountered limestone, and so on down the hole. To determine the composition of the formations, the driller examined the cuttings that washed up to the surface from the bottom of the hole. If he had doubts about what his eyes saw and his fingers felt, he would taste the cuttings.

If the company had a geologist, he would supervise these functions at all wells being drilled. Yount-Lee had no geologist. As the personnel man who had hired Halbouty had said, "Mr. Yount does all the geologizing for us, and I don't remember that he ever went to school a day in his life."

One of the night drillers was Hannibal Westlake, a tough but kindly man in his mid-fifties. He had grown accustomed to seeing Halbouty around his rig, and had become aware of the young man's obsessive interest in the action. He made it a point to talk to Halbouty occasionally, and learned of Halbouty's educational background.

One night he told Halbouty, "You ought to go to the field office and see Mr. Henry. I'll bet he's got logs on every well that's been drilled around here. I'll bet he'll let you see 'em."

E. U. Henry was a civil engineer, chief engineer for Yount-Lee. The next day Halbouty went to the field office on his thirty-minute lunch break. He was fortunate to find Henry there because the engineer spent most of his time in Yount-Lee's Beaumont office.

Henry was a short, stocky, gray-haired man who chain-smoked Picayune cigarettes. He peered through slitted lids as Halbouty explained who he was and what he wanted. He didn't interrupt or attempt to stem Halbouty's spate of words. He merely jerked his head at a large filing cabinet.

"Take what you want but be sure you put them back," he said, as if such requests were commonplace.

Halbouty loaded all the maps and logs he could carry into his Ford jalopy. After work he took them to the Berwick Hotel in High Island, a two-story boardinghouse where the surveying crew was lodged. He began studying in his room immediately. The landlady's teenage daughter, Ruby Berwick, was happy to bring his supper to his room. She thought she had never seen a young man so interesting and so handsome.

Halbouty didn't touch his supper that first night. He studied the maps and logs. The next evening he bought supplies in the village and in his room began "plotting logs" as he had learned to do in college. His aim was to correlate the formations in the wells in which he was interested, those already drilled or being drilled on leases obtained from the Cade estate.

He hung narrow strips of paper on his wall, each strip representing a well that stood in the line of march from the farthest reach of the flank to where the salt shaft thrust through the formations. He ruled off the strips in one-inch segments, each inch representing one hundred feet of well depth. On the strip representing the outermost well he marked off the thickness of each distinct formation the drill had pierced. Then he colored each formation a different hue with crayons—red for shale, for example, green for limestone, blue for sand. He did the same thing on the other strips.

Because they had been uplifted by the salt shaft, the formations were higher near the shaft, deeper on the flanks. A piece of twine strung from a shale formation in the outermost well hung in a rising arc across the strips until it reached the same shale formation in the strip nearest the salt shaft.

Halbouty's studies and his common sense had prepared him to expect this. Nevertheless, he was puzzled. He discussed his puzzlement with Ruby Berwick. "The formations just don't rise high enough near the salt shaft," he said. "Out on the flank the rise is gradual, as it should be, but it stays that way all the way to the salt. It shouldn't do that. That string should *really* climb when it gets close to the salt."

He went to the books he had brought with him from school. They were no help. The Spindletop salt shaft looked roughly like a barber's pole in the textbook illustrations. And the formations rose more precipitously as they approached it. Why didn't the formations do the same at High Island?

He did not reason the answer; it came to him in an intuitive flash. As he stared at the plotted strips on a weary midnight he suddenly *knew* that the formations did not rise sharply where he expected them to do so *because the salt shaft was not where he had expected it to be!*

If the Spindletop salt shaft resembled a barber's pole, the High Island salt was shaped like a mushroom, a toadstool! He was convinced that the salt shaft in its upward journey had struck a resistant layer of sediments, which had caused it to flow horizontally before the resistant layer was penetrated.

This meant that the formations rose precipitously under the mushroom's overhang—and that's where the oil would be if oil there was!

Should he tell someone what he had deduced? His first impulse was to do so, at that very moment. But who would he tell? Who would listen to him and grasp what he was saying? He slumped on his bed in frustration. Then he was back on his feet. He rustled through the maps showing where wells were being drilled or had been planned on the Cade estate tract.

Some were too far out on the flank. Some were on the crest. Only one appeared to be located at a spot where the drill could penetrate the overhang. It was the Cade 21, so named because it was the twenty-first well to be drilled on the Cade tract.

The next evening after his chain-pulling stint he hurried to the Cade 21 drill site. A core had just been pulled and laid out on the rig floor. Halbouty bellied up to the rig. The core was white as snow, about ten feet of pure salt.

But on the bottom end of the core Halbouty thought he saw a shadow, a trace of sediment. At that moment he heard the toolpusher say, "All right, boys. That makes sixty feet of salt we've cored. Tear the damned rig down and let's move it to the next location." The toolpusher was the straw boss, supervising the Yount-Lee drillers at High Island.

Halbouty clambered up on the rig floor and grabbed the toolpusher's arm. "You can't tear this rig down! You've got to take another core!" He pointed to the core on the floor. "You've gone through the salt!"

The toolpusher, Dad Kellam, jerked his arm free. He stared at Halbouty in amazement. "Just who in the hell are you?" he demanded.

"I'm Mike Halbouty."

"Get your ass off this rig, Mike Halbouty, or I'll throw it off!"

"But you're through the fucking salt!" Halbouty screamed.

Kellam shoved Halbouty off the rig floor. He looked down at the young man sprawled on the ground. "I've got orders from Miles Frank Yount to drill into the salt until I can see that it's solid, then skid the rig. If you want to argue about it, argue with Mr. Yount." He turned away from Halbouty. "Tear it down, boys."

In the litter where he had fallen, Halbouty found a dirty ice cream carton. He reached up on the rig floor, tore a chunk off the bottom of the core, and tucked it into the carton.

Kellam had turned around in time to see him do it. He shook his head as if to say, "How did this punk get in the oil patch?"

Halbouty ran to his jalopy and headed for Beaumont, fifty miles away. He knew that Yount lived in a splendid mansion on Calder Avenue, though he had never seen it. Nor had he ever seen Yount, not even from a distance. But Yount was the richest man in town and *everybody* knew where he lived.

It was almost nine o'clock at night when Halbouty reached the Yount neighborhood. The mansion was ablaze with lights and so was the fenced-in backyard. Late-model cars were parked along Calder Avenue for several blocks and on side streets as well. As Halbouty drove slowly toward the residence he was halted by a security guard, one of several he saw at various vantage points.

The guard looked at the old Ford, the young man in stained khakis and muddy boots. "What the hell do you think you're doing around here?"

"I work for Mr. Yount and I've got an important message for him from the High Island field," Halbouty said with aplomb.

"Why didn't somebody call him on the phone?" the guard asked suspiciously.

Halbouty hadn't thought about phoning Yount, but he said, "We tried to call him from the field office but something was wrong with the phone. Something's always the matter with that damned phone." Now Halbouty showed impatience. "I've got to see him. It won't wait."

"All right," the guard said reluctantly. "Follow me and I'll get you parked up close to that big gate."

Halbouty allowed himself and his jalopy to be escorted to a parking place near the vendors' entrance, which also was guarded by a man in uniform. As he walked up to the gate he could hear someone playing the piano on the other side of the fence. He nodded appreciatively as he recognized the melody of *The Moonlight Sonata.* Halbouty came from a musical family; a younger brother was an accomplished violinist.

The gate guard had seen his fellow worker lead Halbouty to the parking place, so he was easier to deal with. He agreed to see what he could do when Halbouty told him what he wanted. "What's going on in there?" Halbouty asked him.

"Mr. Yount's giving a party for some big shot from across the pond," the guard said. "Named Paderewski or something like that. Ever hear of him?"

"Sure. He's a great pianist . . . from Poland. Used to be the president or prime minister or something like that. That must be him playing the piano in there."

The guard was impressed. "Let me see if I can get you inside." He returned shortly with a butler, splendid in livery and haughty as a hotel clerk with a full house.

Halbouty told the butler he had to see Yount. The butler observed Halbouty's dress. "I think not," he said crisply.

Halbouty moved forward and said, "It'll be your ass if you don't let me see him! He'll lose millions because of you!"

The butler flinched and slammed the gate in Halbouty's face, certain that he had repelled a madman.

Halbouty stalked away cursing, seeking a new entry route, but he stopped when he saw a catering truck pull up to the gate. While the guard talked to the truck driver, Halbouty sneaked to the rear of the truck, opened the doors, and climbed inside among boxes and cases of refreshments. The truck rolled into the compound. When it stopped, Halbouty flung open the doors, got out, and looked around him.

He had never seen such opulence except in a motion picture. Dozens of men and women were seated in a garden under colored lights, their array as colorful as the flowers that bloomed around them. The women wore gowns that looked like silk to Halbouty, and the men were in full dress or dark suits. Long tables with snow-white coverings were laden with great bowls of shrimp, steaming roasts of beef, smoked turkeys. On smaller tables punch bowls squatted. A small group of men and women stood around a piano on a raised platform, shielding the player from Halbouty's view.

Halbouty came out of his trance when the butler appeared in his field of vision. The butler spotted Halbouty and moved quickly to a slender, handsome man in the group around the piano. Halbouty deduced that the slender man was Yount. Still clutching the ice cream carton, he hurried toward the man, talking as he went, trying to explain his presence.

Yount was reputed to be a warm, affable man, gracious to the point of diffidence, but now his face tightened in anger as the raucous youth, waving the carton wildly, bore down on him. All conversation among the guests ceased as they stared at Halbouty, but Paderewski continued playing as if he were alone in a rehearsal hall.

"Get out of here!" Yount commanded. "Right now!"

Halbouty stood his ground. Words poured out of him, but Yount would not listen. The butler went after the gate guard. Halbouty still talked, resisting Yount's pushing hand. Then a Paderewski aide touched Yount on the shoulder. "The premier says to let the young man finish so he can finish," the aide said.

Yount immediately was abashed at his own conduct. He took Halbouty's arm and led him to the back door of the house. Paderewski continued playing. The butler approached with the gate guard but Yount waved them away.

"Calm down," he said to Halbouty. "What are you trying to tell me?"

"I work for you," Halbouty said. "At High Island. I know where the oil is."

"What do you do?" Yount asked.

Halbouty told him, then hurried on. "I'm a geologist, Mr. Yount. The salt dome is a mushroom," he said, fashioning one with his hands. "They've drilled through the salt on

Cade Twenty-one but they don't know it. They're skidding the rig."

Yount interrupted. "How do you know this?"

Halbouty showed him the chunk of core in the ice cream carton. "That's sediment," he said, touching the dark shadow with a forefinger. "They wouldn't listen to me." He grasped Yount by the arm. "One more core will prove what I'm saying! One more! I'll stake my job on it!"

Yount's natural good humor was restored by Halbouty's last remark. "All right," he told his lowest-paid employee, "if you're willing to gamble your job, I'm willing to stand the expense of another core. You wait here while I call the field office."

He went into the house. Halbouty waited, turning to face the garden and the people there. Paderewski was still screened from Halbouty's view. The others were busily ignoring Halbouty, acting as if the intrusion had not occurred.

Yount returned. "Get back to the field, son. They're going to take another core. I want you to call me tomorrow and let me know what happens. Tell Kellam I want you to make the report."

He escorted Halbouty to the gate, then walked with him to the jalopy. "Be careful driving back, son." Halbouty got behind the steering wheel . . . and the jalopy wouldn't start.

"Take it easy," Yount said. "You've got it flooded."

"It's not flooded, damn it," Halbouty said contentiously, and he continued to try to start the jalopy.

Yount sighed, rolled the starched cuffs back over the sleeves of his swallowtail coat, and lifted the jalopy's hood. He began tinkering with the engine. "Something's wrong with the magneto," he muttered. "Now try it," he called to Halbouty.

Halbouty tried it, and the engine came alive. Yount lowered the hood and stepped back on the curb, his hands black with old grease. Halbouty pulled away from the curb, twisting in the seat to wave goodbye to Yount. Yount was shaking his head.

For all his youthful cockiness, Halbouty suffered some misgiving as he coaxed the jalopy in the direction of High Island. The nearer he drew to Cade 21 the clearer it became that he would be laughed out of the company if he was wrong, even if Yount asked him to stay on the payroll.

But he swallowed his uneasiness as he parked the jalopy and went to the rig. It had been reassembled and the crew, under Kellam's direction, was preparing to pull a core. Kellam snorted when Halbouty got up on the rig floor. "Welcome back, genius," he said sarcastically. "You've cost the company a lot of time and money, and you've made Miles Yount act like a damned fool."

"Your job is to take cores, and my job is to report what we find to Mr. Yount," Halbouty said evenly. "And don't you ever lay your hands on me again." He sat down on a bench to wait. At his hand lay the driller's log. He picked it up.

The bit had hit the salt at 4,788 feet and had drilled into 60 feet of it before Kellam had ordered that the hole be abandoned. Coring, then, had been resumed at 4,848 feet.

Halbouty became aware that a good-sized crowd had gathered around the rig. Word of his confrontation with Kellam and his wild dash to Beaumont apparently had spread around the field. The crowd was on hand to witness an upstart's humiliation. He placed the log on the bench and composed himself to wait.

It was still short of daylight when the fifteen-foot core was laid out on the rig floor. "Well, I'll be a son of a bitch!" Kellam grunted, and the others milled around him. Halbouty pushed forward to have his look.

The core contained salt, in stringers and parcels, but its main body was gray shale. Kellam turned to Halbouty. "You don't have to say you told me so. I'll say it myself. You were as right as rain." Before Halbouty could reply, Kellam ordered another core to be pulled. He had to shout now to make himself heard above the excitement around him and, old professional though he was, he couldn't smooth the tremor in his voice.

The drill pipe with the core-barrel assembly was lowered back into the hole. When it reached the bottom, coring was resumed. Almost immediately the driller told Kellam he thought he was cutting into a sand formation; there was very little resistance to the core bit.

This *was* exciting news. When the assembly was packed it was pulled to the surface. Kellam waved the onlookers back while the assembly was opened. Halbouty had stood like a statue all through the coring operation, but his stomach was burning with acid. Now Kellam motioned for Halbouty to join him in inspecting the assembly's contents.

What Halbouty saw was a shimmering column of oil-saturated sand. His fingers went out and touched it. He fought an impulse to crush the core in both hands to see the oil drip from it.

He looked around in wonder at Kellam. Kellam nodded knowingly. "There's nothing else like it in the world, kid. I've seen a thousand, I guess, but it still gets me every time." He grinned at Halbouty. "Now let's be hoggish and see if we can find some more."

Kellam guessed that the oil sand was twenty-five or thirty feet thick, and it turned out to be twenty-eight. Several thin sands were encountered down the hole, and at 5,051 feet some twenty-six more feet of pay dirt was cored.

Halbouty already had gone to the field office to call Yount about the first good oil sand. He was worn out and his voice was husky. "It's there, Mr. Yount. The prettiest oil sand you ever saw."

"What does Kellam say, son?"

"He says we've got us a real oilfield—and he's right."

"Get some rest and then come see me," Yount said. "You've found what everybody has been looking for."

The meeting the next day in Yount's downtown Beaumont office was decidedly brief, considering the events that had prompted it. Halbouty, spruced up in clean khakis and polished boots, sat in a chair Yount offered him in front of Yount's wide glass-topped desk. Yount looked across the desk and asked, "How much money do you make with us?"

"Eighty dollars a month," Halbouty said.

"Well," said Yount, "you're making a hundred and fifty now, and it's retroactive to your first day on the job. And from now on you're my geologist and petroleum engineer. I want you ready to begin work with me tomorrow. We've got some drilling sites to stake. How does that set with you?"

Halbouty said evenly, "I want to help them complete the well."

Yount stared at him in disbelief. "That's not your kind of work. You're a geologist."

"I want to help complete it, Mr. Yount."

Yount's exasperation was chased away by a quick understanding of Halbouty's thinking. Cade 21 had become *Halbouty's* well, Yount realized. Of course he would want to see it through to completion.

"All right," Yount said. "When you're through, report to me."

They shook hands across the table and the meeting was over. But as Halbouty was going through the office door, Yount said, "Spend some of that retroactive money as a down payment on a new car, Mike. That flivver of yours is going to fly to pieces."

Almost all the hard work on Cade 21 had been done. The well had been "spudded in" on April 20, Halbouty had begun dragging his chain in early June, and now it was mid-July.

The term "spudding in"—commencement of drilling—was a hangover from the early days. On the old-style rigs a tool combining the characteristics of a spade and a chisel was twisted into the ground under heavy weight, making it gouge out the well. The tool was called a spud.

Cade 21 was drilled with "fishtail" bits, so-called because that's what they looked like. They came in several sizes. The first 30 feet of the well, for example, was drilled with a bit that cut a hole 15 inches in diameter. This wide hole accommodated the 13-inch "surface pipe," which was cemented in the hole to prevent cratering. A device called a "blowout preventer" was locked at the top of the surface pipe. If the well threatened to blow out of control during drilling, the device could be closed to hold back the pressure. Sometimes they didn't work.

Drilling rigs were in operation around the clock once the well was spudded in. In that Depression time two crews worked on the rig, one from 7:00 a.m. until 7:00 p.m., when the second crew took over for its twelve-hour stint. A crew consisted of a driller and four helpers called "roughnecks." Roughneck was not a derogatory appellation. Indeed, the job required strength, stamina, and coordination between mind and muscle in greater measure than did most other oilfield jobs.

The driller was the "prince of labor" in the oilfields. It was he who sent the bit and heavy drill pipe into the earth. He was conscious of every strain on his equipment, every stress on the derrick, feeling his way through the different strata like a blind but clever seamstress stitching an intricate pattern.

The drill pipe to which the bit was attached came in 30-foot joints, 4½ inches in diameter. They were heavy, and the deeper the well became, the greater the weight and strain that was placed on the derrick. The pipe was hollow, and chemically treated "holding mud" was pumped down it to emerge through holes in the bit. The mud tended to keep the bit cool; it washed the cuttings to the surface and cleared the way for the bit to continue boring; it helped prevent down-hole caving by consolidating loose formations such as sand and gravel; and it offered resistance to unexpected encounters with gas under pressure.

A "rotary table" in the center of the rig floor spun like a top at the driller's bidding, turning the pipe and bit in the hole. When one joint was drilled down, another was added to the "string."

Studying the Cade 21 log, Halbouty saw that below the surface pipe a 12-inch bit had drilled 1,000 feet, and 1,000 feet of 10-inch "casing" had been cemented in. Casing was what the word implied—pipe that prevented the well walls from caving in and that would serve as a conduit, if necessary, for the anticipated hydrocarbons.

The remainder of the well's 5,077-foot total depth was drilled with a 9½-inch bit, and a 7-inch casing was cemented in. Halbouty worked as a roughneck, helping set the casing. The cement was pumped down the casing and forced back toward the surface between the casing and the earthen walls. Then Halbouty helped "bring the well in."

Even heavier mud was pumped down the hole. Then a device with several names but generally called a "casing slasher" was lowered into the hole. At the 28-foot interval of oil sand, the device was triggered. It cut ½-inch slits in the casing. It was lowered to the 26-foot interval below 5,051 feet to slit the casing there.

Clear water was pumped down the hole through a 2-inch tubing. The water washed the holding mud out of the hole and the oil and gas were free to enter the casing through the slits made by the casing slasher.

Meanwhile, an assortment of valves—called a "Christmas tree" because it resembled one—had been placed on top of the surface pipe. Then the tubing was sealed to the casing with a "packer," which effectively blocked off the casing as a conduit and permitted the oil and gas to flow only through the tubing.

The oil, under pressure from its accompanying gas, surged toward the surface through the tubing. It was allowed to flow through a $\frac{1}{4}$-inch aperture, or "choke."

The flow was measured at 700 barrels of oil per day. Had the oil been allowed to flow through a larger aperture or the 2-inch tubing, it would have "gushed" into the derrick at about 4,000 barrels per day.

"It's a damned fine well," Dad Kellam told Halbouty when the job was done. "I think you and me deserve a cup of coffee in that new cafe in High Island."

Halbouty was grateful for the respite. He was wearier from the tension than from the labor. At the first flow of oil into the gauging pit he had not been able to restrain himself; he had put a hand into the flow to assure himself that it really was oil, and then he had hidden his triumph by casually wiping his hand on his trousers.

As Halbouty and Kellam entered the cafe they saw Yount at a rear table with two men. Yount smiled broadly and waved at them. One of the men with Yount was young, in khakis; the other was elderly, with a bent back. He wore a white shirt and black trousers held up by wide black suspenders.

Halbouty and Kellam took a table near the front of the cafe and ordered coffee. "You see that old man with Mr. Yount?" Kellam asked. "That's Dad Joiner. Know who he is?"

Halbouty nodded. Columbus Marion (Dad) Joiner the previous year had discovered the greatest oilfield on the continent, the East Texas field, known to the romantic as the "Black Giant."

"He's a real wildcatter," Kellam said. "He don't give a

damn about owning oil. All he cares about is finding it and selling out."

"Mr. Yount's a real wildcatter, too," Halbouty countered. "He goes out to unproven territory and drills. He gambles."

"Oh, sure. He's a wildcatter, all right, but he's also what you call an independent. Yount-Lee is an independent oil company. Mr. Yount wildcats, but he keeps the oil. He produces the oil he finds and he sells it to other people as he produces it, not in one big trade like Dad does."

"He sells it to the major oil companies, doesn't he?" Halbouty asked.

"Most of it. He sure does. But, hell, the majors wildcat, too. They find some oil. Not much, because they've got to count all their fingers and toes before they know who's running the show. It takes a guy with a certain kind of guts to be a wildcatter. Hell, he'll mortgage everything he's got and all he can steal from his grandma's church money to dig a hole a way to hell out in some prairie. You don't find guys like that working for other people. Dad Joiner would die if you stopped him from hunting oil, and so would Mr. Yount."

"The majors do all right," Halbouty said.

"Sure they do. They produce their own oil and buy other people's oil, and then they refine it and sell what they make from it. That's what makes them different from Yount-Lee."

"I'm glad I work for a wildcatter like Mr. Yount," Halbouty said.

Yount and his companions had finished their coffee. As they walked past Kellam and Halbouty on the way to the cash register, Yount paused and put his hand on the check that lay near Halbouty's hand.

"Let me buy your coffee, gentlemen," Yount said.

"No!" Halbouty said quickly.

Yount laughed and said to Kellam, "That's right. It's his big day, isn't it?"

two

It seems as if Mike Halbouty's early years deliberately prepared him for his rendezvous with Cade 21. One of six children born to Lebanese immigrants who ran a small grocery store in Beaumont, he was, in the classic Horatio Alger mold, poor, bright, and hard-working. He sold more newspapers than any other newsboy in town because he earned the best corner in town with his persistent hard-sell, and he could whip any other newsboy who coveted it. He was smart in school and good to his parents. He also was loud, opinionated, and belligerent, and he retained more than a trace of these characteristics through the years.

He got his direction in life while a junior in high school, in studying astronomy—the relationship between the earth, the sun, and the other planets. Halbouty was puzzled and a bit disturbed by the text he was reading, and one day after class he went to the teacher.

"How old is the earth?" he asked. Halbouty was a member of the Eastern Orthodox Church, raised in the faith.

The teacher was cautious. "The earth is very, very old, Mike."

"More than five thousand years old?"

"You'll have to determine that for yourself, Mike."

"Is the Bible wrong?"

She touched his arm. "You'll find the answers to your questions as you get older."

Halbouty didn't want to wait until he got older. He went to the public library—and discovered geology. In the excitement of entering this new world he forgot the points that had troubled him. Many years later he told a friend, "Suddenly I wanted to know all that I could about the earth, the place we live on. I got a kind of spiritual uplift just from thinking that I was going to learn about the earth."

He kept up with his classwork, but he spent as much of his free time as he could in the library, fighting his way through the volumes friendly librarians found for him. And at the same time he went to work in the Spindletop oilfield as a waterboy, carrying buckets of ice water from drilling rig to drilling rig. His pay was fifty cents a day. When summer recess came and he was able to work a full shift, his pay shot up to two dollars a day.

He saved his money and learned what he could about drilling. By the time he was graduated from high school, he knew he wanted to be a petroleum geologist, and he set out for Texas A&M University in College Station, Texas, about 180 miles from Beaumont. He had never been away from home before.

At the school registration office he laid his $125 on the counter. "You're fifty dollars short," the registrar told him. "Sorry. See us next term."

Shocked and downhearted, Halbouty walked around the campus, delaying the journey home. His parents couldn't afford to give him $50, and he knew no one from whom to borrow it. He felt that if he returned home he would spend the rest of his life in the family grocery store.

As he walked about the campus, passing groups of happy young men, he kept hearing a name. "Prexy Walton fixed everything," he heard one youth say. "Prexy Walton can handle it," another said. All around the campus it was Prexy Walton did this, Prexy Walton did that.

The naive teenager from Beaumont had never heard the word "Prexy." He assumed that Prexy Walton was a big man on campus who knew all the tricks and packed considerable clout. He went looking for Prexy Walton. He was surprised to find that Prexy Walton had an office—and a secretary.

"What can I do for you?" the secretary asked.

"Ma'am, I want to see Prexy Walton," Halbouty said.

There was no malice in the secretary's smile. "Prexy is a slang word for president," she said kindly. "Mr. Walton is president of the university."

Halbouty blushed. "I'm sorry, ma'am, but I still sure would like to see him."

"What about?"

"Well, I've worked hard trying to save money to come here to school, and they tell me I'm fifty dollars short. If I go back home, ma'am, I'll never get another chance to go to school. I can't go back home. So I've got to see him."

A man's voice interrupted him. "Come in this office and talk to me." Halbouty turned and a man was beckoning him from a doorway. "I'm T. O. Walton," the man said.

Halbouty walked into the office. "I heard your story," Walton said. "What do you want of me?"

"I want you to lend me fifty dollars," Halbouty said. "I guarantee that I'll work my shirt off to pay it back."

Walton took his wallet from his pocket. He thumbed through his bills and handed Halbouty two twenties and a ten. "Go register," he said. Halbouty started to thank him, but Walton held up a hand to silence him. "Go register."

Halbouty registered. He didn't have as much as a penny left in his pockets, but registration fees in those days covered tuition, room and board, and even tickets to home sports events. And the next morning he found a job—mowing lawns for twenty-five cents an hour, the going rate. He worked after school hours and on Saturdays and Sundays. He ran errands for upperclassmen, for pay. He paid his debt to President Walton before the term was out.

During the summers Halbouty worked as a laborer in the Magnolia Refinery at Beaumont or as a truck swamper in the Spindletop field. He pushed a lawnmower and did odd jobs during the school term. There was no time for a vacation.

But he was a distinguished student, with a list of honors and accomplishments as long as a joint of drill pipe. (Courses in petroleum engineering were introduced in the school when Halbouty was a sophomore. The teachers were well grounded in theory, but had little or no practical experience. Halbouty was selected to lecture on the basics of oil-well drilling, something he had picked up as a waterboy and truck swamper at Spindletop.) He also was active in the Corps, the school's famed ROTC unit.

His aptitudes made him the favorite of the head of the school's geological department, Dr. John T. Lonsdale. When Halbouty obtained his Bachelor of Science degree in 1930, Lonsdale had a proposition for him. He wanted Halbouty to get his Master of Science degree, and he knew how it could be accomplished.

The United States Geological Survey and the State of Texas had engaged Lonsdale to make a hydrological study of a rich watershed in the Winter Garden area of Southwest Texas. Lonsdale wanted an assistant. He coaxed the State Legislature into financing a fellowship for graduate work— the only such fellowship on record—and the school bestowed the fellowship on Halbouty, to act as Lonsdale's assistant. The grant was for $750, paid in advance, a bonanza for a youth who had pinched pennies for four years.

Lonsdale gave Halbouty the task of working out the geology of Atascosa County. Halbouty's report would constitute his thesis for his Master's degree. Free of the classroom and campus restrictions, ready to apply his learning for the first time in the field, Halbouty ranged over Atascosa County with such enthusiasm that the sheriff complained to Lonsdale. "That son of a bitch is worse than a Peeping Tom," the sheriff griped. "Every time I go out on patrol I see him running across somebody's ranch or digging around in somebody's field. Somebody's gonna pepper his ass with buckshot if you don't slow him down some." Lonsdale promised to calm down Halbouty, and he did.

Halbouty's report—his thesis—was impressive. In working out the geology of the county he had made the first base map of what would be called the Charlotte-Jourdanton-Leming fault system. And he had an additional word for Lonsdale. "I think there are some good oil prospects in here," he told Lonsdale. "Why don't we go in with somebody to lease 'em and drill 'em?"

Lonsdale, a pure scientist, shook his head. "Ah, Mike. You know I'm not interested in things like that. Besides, I've got other plans for you. So forget it."

Looking back over the years one can hardly believe that an aggressive, cocksure young man like Halbouty would bow so readily to Lonsdale's admonition, but he left Atascosa County without a backward glance. It is stranger still that he didn't make a foray into Atascosa County after he "discovered" the High Island field. Certainly Yount would have been receptive to any recommendation from his "boy genius." But there is nothing on the record to suggest that Halbouty ever mentioned the Atascosa County area as an oil prospect to anyone but Lonsdale.

It was a major mistake. Some years later, Humble Oil & Refining Company led the way into Atascosa County and found the first of fifty-three oilfields. A company geologist had obtained a copy of Halbouty's thesis from the university library, and the company used it as a guide to the liquid treasure along the Charlotte-Jourdanton-Leming fault line. The county has produced more than 112,000,000 barrels of oil and not a one belonged to Halbouty.

Halbouty had three ways to go when he received his Master's degree. On Lonsdale's recommendation, the United States Geological Survey offered Halbouty a job as a hydrologist in Oregon at $150 a month. Lonsdale also used his considerable influence to obtain Halbouty a $1,500 fellowship for a doctorate at the University of Missouri. And the United States Army Air Corps, impressed by his scholastic record and his ROTC accomplishments, offered him an appointment as an Air Cadet at Brooks Field.

Halbouty leaned toward the University of Missouri fellowship, and Lonsdale urged him to accept it, but he was eager to get into the oil business. The activist in him was at war with the scholar. So instead of accepting either offer, he took the first oilfield job he could land—chain-puller for Yount-Lee

It was well he did. Miles Frank Yount took him under his wing, and Yount was an oilman without peer.

Though Cade 21 was the first well to find High Island's abundant supply of crude under the mushroom overhang, it was not officially recognized as the field discovery well. That honor went to a well drilled on the crest of the dome some nine years earlier, a piddling producer that spouted fitfully and died. Other wells drilled through the overhang found richer sands, some more than two hundred feet thick. And later on other companies and wildcatters found producing sands by drilling deeper wells on the flanks.

The producing sand could not be immediately identified as one of the recognized oil-bearing strata. Generally oil sands were named for the community or area where they outcropped at the surface. The Woodbine sand, for example, was named for a small North Texas community near Dallas where it was first hailed as a prolific water-bearing stratum. At High Island people talked about the "Cade sand," and that became its name.

Cade 21 was drilled at a time of over-production and falling prices in the oil business. The Depression and vast imports of foreign oil by major companies with overseas fields had, by 1930, pushed down the price of domestic crude from $2.29 a barrel to $1.10. Then, in October 1930, eight months before Halbouty joined Yount-Lee, Dad Joiner discovered the East Texas field, whose reservoir contained more than 6 billion barrels of high grade crude. By February 1931, with wells roaring in around the clock in the new field, crude was selling for fifty cents a barrel.

A bill to tax imports heavily was defeated in Congress; the majors had the political clout. State orders to limit production in the East Texas and other Texas fields were overturned in court or ignored in the fields. On August 15, some eleven weeks after Halbouty signed on with Yount-Lee, it was announced that the East Texas field was producing a million barrels of oil a day! And the next day the Texas governor declared martial law in East Texas and sent in the National Guard to enforce it. For the most part, the oilmen outwitted the "Boy Scouts," as they called the guardsmen, producing oil to the limits of a well's capability; many months would pass before the Texas Legislature would enact meaningful proration laws, and see them enforced.

Yount was deeply involved in the East Texas field, and he was among the leaders in the fight for sane exploitation of the giant reservoir. He had paid $3,270,000 for the third well drilled in the field and the five thousand acres on which it sat. When his public cries for responsible drilling went unheeded, he worked behind the scenes with others for effective legislation. And he saw to it that his own company drilled wells with proper spacing and respect for the fragile reservoir.

With proration, prices began to rise slowly. But even with depressed prices, those who produced oil were the Arabs of the 1930s. It cost less than twenty-five thousand dollars to drill a five-thousand-foot well, and many were completed in fifteen days or less. A good well could pay out its cost in a month. There was virtually no income tax. And in 1926, when the oil industry convinced Congress an oil shortage was imminent, the legislators granted the industry the Depletion Allowance, which permitted the industry to pocket the first $27\frac{1}{2}$ percent of gross income. The untaxed money was to be spent in exploring for new oil, and many times it was.

For major companies with overseas holdings it was even better. The foreign oil could be produced at a fraction of the U.S. cost. In Venezuela, for example, the government received only a 7 percent royalty on the oil the companies shipped to the United States and around the world. And Venezuelan oil could be shipped to the Eastern Seaboard more cheaply than Texas and Louisiana crude could be shipped from Gulf of Mexico ports.

Halbouty knew little about oil economics and politics, and cared less. The rich core at Cade 21 had made him a discoverer, a man apart, and all he wanted to do was to find more.

It would be said in petroleum circles that Halbouty never selected a drill site for Yount-Lee that the drillers didn't find oil. If true, it was because of Yount's restraining, guiding hand. Yount already had gained a reputation as a man who discovered oil in most of the wildcat areas he entered. His uncanny success led some to believe that he used some kind of "wiggle stick" or "doodle bug." The truth was that he was a self-educated geologist, a "creekologist," he called it, and he had the courage to play his hunches. He had left school in the third grade, but through reading and studying he had become a man of parts, a collector of fine art and a lover of music and literature.

It was a mark of Yount's sensitivity that he did not tell Halbouty that he, too, had deduced that the High Island salt dome was a "mushroom." And that the man from whom he had acquired most of the High Island acreage, Marrs McLean, had arrived at the same conclusion. Halbouty would learn it from others, years later, and he would pay homage to both in ways open to him.

During the second week after Halbouty's elevation to company geologist and petroleum engineer, the young man rushed excitedly into Yount's office to display a core taken from a well in a relatively new area of High Island. The core was saturated with oil. Halbouty wanted to begin drilling immediately to the east of the well from which the core had come. "We can get another oil well," he told Yount.

Yount leaned back in his chair. "Mike, I don't think you have any idea yet of how faulted a salt dome is."

"Oh, I know," Halbouty said airily.

Yount smiled. "No, you don't. You haven't had the experience to know. When that salt shaft rises, it absolutely shatters formations."

"I know that," Halbouty said.

Yount pointed to the core. "There may be a dozen faults between this well and the one you want to drill. You can't set a pattern like you can in other oilfields. Every well you drill at a salt dome is a wildcat."

Halbouty's nod lacked total conviction.

Suddenly Yount bent over and wrenched off a shoe. Holding it by the toe, he brought the heel down solidly on the glass topping of his desk. The glass cracked in hundreds of wavy lines.

"Now," Yount said to the startled Halbouty, "*that's* what you look at when you look at a salt dome. Do you understand?"

"I sure as hell do," Halbouty said.

"Don't forget it," Yount said. He put on his shoe.

While Yount was educating him in "creekology," Halbouty was getting another kind of education from Yount-Lee employees. Many of them resented his status with Yount, his success in facing down Dad Kellam. And they considered him an arrogant prick, as devoid of humor as a pound of calf liver. Drilling crews tried to outdo each other in playing tricks on him, and they were never delicate.

Halbouty was lost in his work, and he considered such japery a mere hindrance. He took the rawhiding stoically; had he belly-laughed one time he might have found peace. The time came, of course, when he was goaded into combat.

The roughnecks had derisively nicknamed him "Mud Eater" because of his method of examining a core taken from a well. After recording the composition of the core, Halbouty would taste it in several areas, looking for oil signs. Then he would smell it. It was a never-failing routine, and the roughnecks soon became aware that he tasted before he smelled.

And so it was one morning that he tasted an area in a core that looked much like lime. His tongue—and the roars from the watching roughnecks—told him it was sun-dried dog excrement. Halbouty's smile was weak and his laughter unequivocally false.

"We're changing your name to Shit Eater!" cried a driller called Corsicana Slim. He began unbuckling his belt. "Come here and I'll shit in your face." He basked in the admiring laughter.

In the past, Halbouty had walked away. But no one had ever gone this far before. Now he said, "If you shit in my face, the undertaker will wipe your ass, you son of a bitch." He advanced on Corsicana Slim. Joyfully, Slim went to meet him.

They fought for more than half an hour. They fought on the rig floor and on the hard ground. Slim was strong and tough and quick as rain, veteran of dozens of honky-tonk brawls. If there was malice in him, or hatred for Halbouty, it was overshadowed by his love of combat. He fought simply for the hell of it. Halbouty fought out of a cold fury, a proud man intent on whitewashing his humiliation. There was no fun in it for him, as there was for Slim. He meant to win or die.

They fought until there was no more fun in it for Slim. Their faces were bloody and swollen, their bodies aching and quick to the touch, when Slim finally recognized the demon in Halbouty; there would be no hand-shaking, beer-drinking making-up when this fight was over. He tore himself from Halbouty's grasp, ran to the rig, and picked up a twenty-four-inch Stilson wrench, an all-purpose tool also good for cracking skulls and breaking bones. He swung the wrench and hit Halbouty in the part in his hair.

Dad Kellam fired Corsicana Slim, not for fighting but for resorting to the wrench. Halbouty spent eight days in a hospital with a hairline fracture of the skull. Yount acted as if he had not heard of the fight or missed Halbouty's presence in the field.

Halbouty returned to work to find that the fight had won him no measurable favor among the roughnecks and drillers. His ferocity, however, had earned him respect of sorts. While the tricks continued, they were gang affairs; no individual put himself in the forefront as Corsicana Slim had done.

Halbouty didn't develop a sense of humor overnight, nor

did he make any obvious attempt to hide his brilliance. But after long months something occurred that brought him company-wide admiration. Halbouty did not react as if this new attitude was his due. Some hard-won wisdom prompted him to accept the warmth extended him with an unsuspected graciousness.

Cade 30-A "blew out" at mid-morning, coming in with a roar that sounded like the passage of a thousand freight trains. Oil, gas, mud, and salt water spewed high into the slaty sky. The hole spouted sand in belching bursts also, which cut into the wooden derrick like an army of voracious termites.

Yount was drinking coffee in the High Island field office when word of the blowout came in. With him was Dale Cheesman of Shell Oil Company, who had driven down from the East Texas field to see what Yount-Lee was doing at High Island. "She's a goin' Jesse, Mr. Yount," said the messenger from the rig. "She's blowin' out everything but peanut butter and jelly."

"Anybody hurt?" Yount asked quickly.

"No, sir. We all got off that floor like it was being raided."

Yount and Cheesman drove to the scene. The driller and his roughnecks were surrounded by other workmen at the high-and-dry end of the catwalk. The well had settled down to a sullen roar, occasionally punctuated by a sharp rattle as rocks ricocheted out of the casing.

"Them rocks could set the booger on fire, Mr. Yount," the driller said apologetically. "That's why we left there so fast."

Yount nodded that he understood. Halbouty had arrived at the scene in time to hear the driller's remarks. He stood by Cheesman and, like the others, took in the awesome sight. Yount finally shook his head as if he had arrived at a decision.

"Somebody is going to have to go out there and find out what the damage is so we'll know what we have to do to shut her in," Yount said. He was asking for a volunteer.

Cheesman stirred uneasily, remembering what the driller had said about the rocks. (Many years later he said, "I wouldn't have gone out there if Yount had promised me the whole oilfield.") None of the crewmen stirred, and Cheesman was surprised when the young man standing next to him said to Yount, "I'll take a look."

"Do you know how dangerous it is?" Yount asked.

"There's too much salt water coming out for it to catch on fire," Halbouty said.

"Go ahead," Yount said.

Halbouty set out on the catwalk. The crowd watched him in silence. He reached the rig. He got out his handkerchief and tied it around his face like a western bandit's mask. Then he disappeared under the rig floor.

"I had thought he would dart right under the rig and dart right out," Cheesman said, "but he didn't come out. We waited and waited."

The crowd was growing restless. Cheesman heard Yount muttering to himself. The driller pushed through the crowd to the catwalk.

"I'm going down there," the driller said.

But Cheesman cried out, "Here he comes!" He had caught a glimpse of khaki movement at the edge of the rig floor.

Halbouty emerged. He trotted back on the catwalk, whipping off his handkerchief mask before he reached the crowd.

Discussing the incident later, Cheesman noted, "He had a small notebook and a pencil in his hand. He hadn't gone down there just to take a look and depend on his memory. He had taken time to *draw a picture* of the damage!"

Halbouty was faint and choking when he reached Yount's side; the mask had done little to keep him from inhaling the gas. He handed Yount the notebook, then sat down on the catwalk.

"Give him room," the driller said. He snatched a thermos bottle from a roughneck's hand, opened it and handed it to Halbouty. Halbouty drank.

Yount, meanwhile, was studying Halbouty's drawing. Cheesman looked over his shoulder. The blowout preventer and a shut-off valve were damaged beyond repair, and the oily mixture was shooting out of a ten-inch opening.

Yount summoned the driller and his roughnecks to his side. "You're going to have to stab it," he said. He showed the men Halbouty's drawing. "Take all the help you need," he told the driller.

A "stabber" was put together hastily. It was a short section of pipe with an open valve assembly on one end. The other end was beveled so that it would slip into the pipe through which the oily mixture was shooting. The device was dangled over the pipe. Roughnecks gently moved it into the stream until the fluid was shooting up the stabber pipe and through the open valve. The device was lowered slowly into the pipe and clamped in place. Then the open valve was closed slowly—and the flow was shut off and the well was under control.

"I'll buy the coffee," Yount said to Cheesman and Halbouty, and he drove them to a roadside cafe outside High Island.

As Cheesman recalled it, "Yount was mighty pleased with Halbouty. He kept on talking in the cafe about how the others weren't afraid to go to the rig once Halbouty had showed there was little danger.

"But about that time I heard a clicking. Halbouty was trying to put his cup down in the saucer and his hand was shaking so bad I thought he was going to drop the cup. Yount appeared not to notice, but Halbouty said, 'I'm just now getting scared, Mr. Yount.'

"Yount just nodded his head and said, 'That's normal, Mike. It'll pass in a minute or two.' He acted like it wasn't important at all. And sure enough, in a few minutes Halbouty was all right. He went ahead and drank his coffee, and we all got to talking about the East Texas field.

"Later in the day, back in the field, Yount got me aside and asked me not to say anything about Halbouty's being scared in the cafe. He said, 'You know how roughnecks are. They'd kid him about it regardless of what they really think about his going to the well.'

"I understood, of course. It was easy to see that Halbouty was real sensitive. All I heard while I was around High Island was that he was as nervy as Jesse James."

Not long after the Cade 30-A blowout, Halbouty won the gratitude of his fellow workers.

In the early days, only water was pumped down drill pipe to aid the drilling. As years passed, however, drillers learned that muddy water often was better as a drilling fluid than clear water, and they also found that some muds were better than others. Clays, in particular, sometimes prevented down-hole caving by consolidating looser formations.

By the time Halbouty arrived at High Island, the major oil companies and a few outside firms were conducting extensive experiments with drilling mud, testing clays and combinations of clays and introducing chemicals into the mixes. Some of the products realized went a long way in reducing blowouts and improving almost every aspect of drilling.

But they could find no formula to help combat a pesky formation that slowed down exploitation of many Gulf Coast fields and threatened to halt exploitation entirely in a few. Oilmen called the formation "heaving shale." It seemed to suck the fluid out of drilling mud. Then the shale would swell and rush up the bore hole. A driller might lift a string of drill pipe from a five-thousand-foot hole to change the bit and find, on going back into the hole, that the shale had filled as much as a thousand feet of it.

Yount-Lee had been fighting a frustrating battle against the shale at Spindletop, Barber's Hill, and some Louisiana fields. Yount was not surprised when the formation was encountered at High Island. The shale slowed drilling in some wells to the point of doubling drilling costs. Some wells were simply abandoned to the shale. That the major oil companies were suffering similarly brought little solace to Yount.

Yount had been planning to build a laboratory. He was selling tankerloads of crude to French interests and paying outside firms to test the crude for constancy. With his own laboratory, he could test the crude himself. It occurred to him, also, that his young chief geologist had "minored" in chemistry at Texas A&M. He took Halbouty for a ride in his sleek black Duesenberg.

"Did you like chemistry in college, Mike?" Yount asked.

"I liked it, and I was damned good in it," Halbouty said. "If I had a place to work, I could whip this damned heaving shale. I could come up with a formula."

Yount braked the Duesenberg to a halt. He told Halbouty of his plan to build a laboratory.

"That'll take too long," Halbouty fretted. "Kellam's already talking about laying off some of our people."

Yount slapped the steering wheel. "Look! We own an old lumberyard in Beaumont. Why don't you set up in it while the lab is being built? Order anything you need." Forgotten was the testing of oil for the French; that could come later.

Halbouty set up shop in the deserted lumberyard. He ordered so many gallon buckets that Beaumont hardware dealers couldn't supply them; Yount sent a truck to Houston to get more. The empty rooms of the lumberyard were filled with buckets, sacks of clay, and bottles and cans of chemicals. Halbouty installed Bunsen burners and other laboratory equipment.

Soon buckets of chemically treated mud were lined up in orderly rows across the floor in the largest room. Halbouty kept meticulous records. He worked far into the nights, mixing various combinations of chemicals with the clay. And he got nowhere. The mixture he sought, one fluid enough to move and strong enough to hold back the shale, eluded him.

Weeks passed. Yount was concerned for Halbouty's health, for the young man had become gaunt-eyed and irritable. One night he found Halbouty asleep on the floor between two rows of buckets. When Yount woke him, Halbouty went back to work. "You can't keep this up much longer," Yount said gently.

"Trial and error," Halbouty said. "It's the only God-damned way. I'll keep on until I find it."

"But you've tried almost everything," Yount reminded him.

"Not the phosphates, Mr. Yount. I'm going into them in the morning."

Four days later, at three o'clock in the morning, Halbouty called Yount at home. "We've got it," he said. "Something to whip the shale."

"I'll be right there," Yount said.

Yount arrived in his house shoes and with his pajama top pushed down in his trousers. Halbouty gave him a cup of coffee, then showed him a pail of mud. The mud looked dense enough to be a solid cake. Halbouty thrust a big spoon into the bucket and stirred. The mud "broke" immediately, attaining fluidity. "Hexametaphosphate," Halbouty said.

"What?"

"Hexametaphosphate." Halbouty spelled the word for Yount. "There are other chemicals in there, but that's the baby that turned the trick." He "broke" the mixture in another pail, and Yount grunted with satisfaction at the sight. "Let's test it right away," Yount said crisply. "We'll try it in High Island first."

Halbouty mixed the mud to be tested, and he stood by during drilling until exhaustion overcame him. He went to sleep in his car, satisfied that the mixture was doing the job. The driller had instructions to waken Halbouty if he encountered any trouble, but it was Dad Kellam who shook him until his eyes opened.

"The driller says the pipe is stuck," Kellam said, "and it damned sure is. That's why I'm here. They called me."

Halbouty came out of the car in a rush. He ran to the rig and mounted the rig floor. "The pipe's stuck," the driller said. "Your mud ain't no good." He seemed more amused than disturbed, and Halbouty saw fleeting smiles on the roughnecks' faces. Halbouty grabbed a bucket and filled it with mud from the slush pit. He went to the back of his car where he had stowed his testing equipment and went to work. Kellam watched him, but neither spoke.

Fifteen minutes later Halbouty strode back to the rig and mounted the floor. Kellam was right behind him. Halbouty looked around at the driller and his crew. "You rotten sons of bitches," he said softly. "You rotten sons of bitches put water in it."

"No, we didn't!" the driller said angrily.

"You did, you lying son of a bitch. You turned the water hose in it. You ruined the mud." Halbouty took a step back. "I'm going to kick the shit out of every one of you."

Kellam stepped in front of Halbouty. "Let me handle this, Mike." He faced the driller and his roughnecks. "Y'all are just about the silliest bastards I've ever seen. If this mud works, if it whips that heaving shale, we're all going to stay on the payroll. If it don't work, a lot of drillers and roughnecks are going to be selling apples on street corners. Now, Mike says you watered the mud, and I believe him. I want to hear the truth from somebody."

"We watered it," the driller said.

"All right. What do you think I ought to do about it?"

"Give us a chance to make up for it," the driller said earnestly. "We'll do anything Mike says to do."

"All right," Kellam said. "I'm passing the word, starting right here, that anybody that tries to fuck with this mud program is going to get run off."

The next day Yount toured the field in his Duesenberg, delivering the same message in more moderate terms.

Halbouty's mixture whipped the heaving shale at High Island and at other fields where Yount-Lee had acreage. It wasn't long before other oil companies noticed that Yount-Lee was completing its wells in record time. Yount called Halbouty into his office.

"I think we ought to patent your formula, Mike," Yount said. "It could mean a lot of money, to you *and* the company."

Halbouty shook his head. "I think we ought to keep it for ourselves, Mr. Yount. We're doing all right the way things are."

"Aren't the other companies trying to get you to tell them the formula?"

"Yes, sir. I've had lots of calls, but I haven't told anyone anything. We don't need a patent."

Yount didn't argue, but he had been too long in the oil business to think the secret would be kept very long. The theft of ideas and information in the oilfield was not considered a crime in those days; indeed, the scout who succeeded in learning details of drilling on a "tight" well, for example, was considered a hero among his brethren and in boardrooms as well.

A Yount-Lee attorney also spoke to Halbouty about patenting the formula, but Halbouty gave him the same answer. Like Yount, the attorney knew that eventually someone would steal a sample from a well being drilled. "They can analyze it as well as you can, Mike," the attorney said. But Halbouty paid him no heed.

There is no record of who first filched and analyzed a sample, but an infant "mud company" brought out a trade name product that whipped the heaving shale. A fifty-gallon drum of it sold for sixty dollars. Halbouty analyzed a sample; it was produced according to his formula.

"How much does it cost us to make a fifty-gallon drum?" Yount asked.

"About thirty cents," Halbouty said.

Yount never mentioned the matter again, but the attorney told Halbouty, "You kissed off a fortune."

If he did kiss off a fortune, Halbouty was rewarded by the warm receptions he received as he traveled through Yount-Lee holdings. He was credited with keeping Yount-Lee employees on the job while across the country men who wanted to work stood in breadlines. If he was still abrupt and cocky, he could be forgiven.

"How are you getting along these days?" Yount asked him as they rode across Spindletop field in the Duesenberg.

Halbouty knew what he meant. "They love me," he said.

Yount smiled.

"They do, damn it!" Halbouty said heatedly.

Yount said nothing. They rode on for perhaps fifteen minutes. Then Halbouty said, "I think a lot of 'em like me, Mr. Yount."

"I'm sure they do, Mike," Yount said.

Halbouty was one of the first "mud engineers," a now-honored profession; none but the rankest of poor-boy wildcats is drilled without a mud engineer in attendance. But Halbouty wanted to extricate himself from the mud, and Yount let him do it. He needed Halbouty's geological expertise in the fields where Yount-Lee was expanding operations.

One of these areas was the East Texas field. It will be recalled that Yount had paid more than three million dollars for a 5,000-acre lease in the field.

The field covered portions of five counties. The reservoir was forty-five miles long and five to twelve miles wide. In all it covered more than 140,000 acres. And there were few dry holes once the limits of the field had been determined. It was only on rare occasions that the bit failed to find the rich oil sand.

But Yount was concerned about conserving oil, a novel idea in those days. "Someday we'll regret every drop we waste," he told Halbouty on one of their long rides together. "See to our interests, see that we do well, but remember that the earth is our home and don't do a thing to desecrate it." He smiled at Halbouty. "And stay out of trouble."

Halbouty was not prepared for East Texas. Kilgore was the center of the boom. One day it had been a rural community of seven hundred. Three days after the second well was drilled in the field, it became home to ten thousand eager strangers. In these communities—Kilgore, Gladewater, Henderson and a dozen hamlets—the greatest boom in world history was taking place.

The only law of any consequence was a fabled Texas Ranger, Lone Wolf Gonzaullas, the most feared and respected man in Texas. Halbouty ran afoul of Lone Wolf his first night in the field. He wandered into Newton Flats, a vast tent city four miles east of Kilgore. The tents covered three acres. The area was devoted to vice, and it did not cater to the effete. Women, booze, and every gambling game known to man were offered to rich and poor alike; a boomer could "shoot a dime" in one dice game while at the adjoining layout a more prosperous citizen could toss a thousand dollars on a single spin of the roulette wheel.

Halbouty wandered into one of the tents and was taken in immediately by a black-haired woman in beach pajamas whose name he never learned. They were on their second drink of popskull whiskey at a pine-slab bar when Lone Wolf sauntered in. There was no mistaking who he was. He wore a

Stetson. He was booted and spurred. There was a badge affixed to his shirt pocket, and two silver-mounted six-shooters were holstered on his thighs.

He made a beeline for Halbouty because Halbouty was a stranger to him. And for the night out, Halbouty had polished his boots. His khaki clothing was clean and pressed; his hair was slicked down and his moustache bristled.

Said Lone Wolf: "Let me see your hands, palms up."

Halbouty had heard the stories: if a man's hands were calloused, he was a workingman and therefore presumed to be honest until he proved himself otherwise; if the palms were smooth, the owner went on Lone Wolf's "trotline" until Lone Wolf ascertained his degree of criminality, if any. Halbouty had thought the stories humorous when he first heard them, but now he found no humor in his predicament. Instead, he was outraged. "I'm a geologist for the Yount-Lee Oil Company. I don't work with my hands."

"Palms up," Lone Wolf said.

"No."

Lone Wolf smiled. "All right," he said mildly, "but I'll have to take you to jail."

Jail, according to the stories, was a vandalized Baptist church in Kilgore. A heavy chain ran the length of the church down the center aisle. From this chain, like ribs on a backbone, about twenty smaller chains ran to both sides of the church. Prisoners were tethered to the smaller chains, which were looped around their necks and padlocked.

"You'd put me on the trotline just for not showing you my hands?" Halbouty asked.

He said later he never observed the movement, but suddenly he was looking into the muzzle of one of the six-shooters. Lone Wolf waved the revolver. "The last time. Palms up, or jail."

"Fuck you," Halbouty said.

He was driven to Kilgore in a panel truck with five other

prisoners. But he was not placed on Lone Wolf's trotline. He had spent little more than an hour in a cell with eight other men when Lone Wolf released him.

Lone Wolf had some words for Halbouty. "If you're here to work, work. Stay out of these gambling hells. And that woman you were with, she's got the clap."

(In 1971 Lone Wolf would tell an interviewer: "There was something about him I liked. At first I thought he was showing off for the woman, and I was ready to split his lips. But he wasn't. He kept saying something about his rights, that I was denying him his, but he didn't understand my problem. I was hard on bunco artists and robbers and killers, and I guess there were thousands of them there. I didn't have time to worry about rights except the rights of the folks who were being taken one way or another or robbed and beaten. But I liked him. I looked in on him several times while he was in East Texas, and I saw him later on in other towns and we would laugh about it all and I kept on reading about him in the papers. He might have made a good Ranger. God knows he was stubborn enough.")

It wasn't Halbouty's only scrape at the East Texas field. One day Gilbert Prince, Yount-Lee's general superintendent, walked into the company's field office to find Halbouty in a terrible cuss fight with Red Herring, the field superintendent. Prince stepped between the two men and demanded an explanation.

A well had been drilled but had yielded only a six-inch core of oil-saturated gravel at the depth where the Woodbine sand normally was from fifty to sixty feet thick. Herring wanted to abandon the well. Halbouty contended the well could be the best Yount-Lee had drilled or ever would drill in the field.

The gravel, Halbouty argued, was still the Woodbine sand; it was coarser because it was at the eastern edge of the field. "I'll bet this well is not more than twenty feet from the limit of the field where the Woodbine pinches out," he said. "If we had drilled twenty feet farther east, we would have missed the sand entirely."

"So what?" Prince demanded. "It's still not much of a core."

"That's what I've been saying," Herring chipped in.

"This is a water-drive field," Halbouty explained. "The water forces the oil through the Woodbine from west to east. Our well, being on the eastern limit of the field, will be producing oil when the wells to the west—those with thick sand—are dead and gone. It damned well might be the last producer in the field!"

Prince thought a moment; he later told friends he was remembering Halbouty's confrontation with Dad Kellam. "Set the pipe, Red, and make us a well," he said decisively to Herring.

The well came in, and it was still producing its allowable in 1978—46 years later—with no hint of a slowdown.

Prince and other ranking Yount-Lee officials were aware of Yount's interest in Halbouty, and some believed Yount was grooming Halbouty for a managerial role in the company. Prince felt that Yount, who had no sons, simply had developed a fatherly affection for the young man. But Dad Kellam said the attachment was stronger than that.

"It's comradeship," he declared one night in a High Island boarding house. "Who else does Mr. Yount have to talk to around here? He gets lonesome for good talk, and the whippersnapper and him like a lot of the same things. Hell, the whippersnapper's teaching him some things, too!"

Kellam was right. Halbouty, after all, was the only technical man on the Yount-Lee staff, and he and Yount talked geology and petroleum engineering for hours on end. They shared a love of the earth, too, and Yount constantly pressed on Halbouty his attitudes about oil conservation and protection of the environment. At one point he changed a drillsite Halbouty had selected. "We don't want to drill here, son," he said. "We would damage these trees." He pointed to a nearby grove of oaks. "There are not many trees on this peninsula, anyway, and we ought to save all we can."

At the same time, Yount seemed to be seeking situations where Halbouty would be exposed to every facet of the company's operations. For example, he stopped Prince and Halbouty on a Beaumont street and took Halbouty through a catechism. Prince told friends the questioning went something like this:

"Mike," Yount asked, "what do you know about pipelines?"

"I don't know anything about them, Mr. Yount."

"Do you mean to tell me you don't know a *thing* about pipelines?"

"Not a thing."

"Good!" Yount said. "We're going to build a pipeline from High Island to Spindletop tank farm, and you're going to be my pipeline engineer. I've got some ideas on how the line ought to be built, and I want you to see that they're carried out!"

So Halbouty became a pipeline engineer on what must have been one of the toughest fifty-mile pipeline projects in history. The inspector on the project was Harry K. Smith, now a Houston industrialist. "It started raining the day we started the pipeline," Smith said. "It doesn't seem possible, but I don't believe it ever stopped raining. We built that pipeline under water. The rain brought out the snakes and it seemed as if someone was bitten every day. And we were using electric welding machines, and someone was always getting shocked. It was a mean job, from start to finish, and if Halbouty didn't know anything about pipelining before, he sure learned."

The contractor, Nick Saigh of San Antonio, was losing his ass and its fixtures on the job, Smith told Yount. Saigh had guaranteed to build the line for a fixed price. When the line was completed, Saigh had spent in construction several hundred thousand dollars more than the price he had given Yount. He was facing economic destruction.

When Saigh went to settle up with Yount, Yount handed him a check for the contract price. Saigh thanked him. "I'm sorry the job took so long, Mr. Yount," Saigh said. "I'm sorry I inconvenienced you."

Yount smiled. He took another check from a folder and handed it to Saigh. "Nick, this check covers the amount of money you overspent, plus a ten percent profit on the job."

There had been no courses in micropaleontology at Texas A&M, so Halbouty began studying every paper he could find dealing with the life of past geological periods as known from fossil remains. The knowledge came easily, and Yount encouraged him to make trips to Houston to learn what he could from major company scientists. Halbouty's ability to absorb information quickly and put it to practical use helped him at this early stage to gain a reputation along the Gulf Coast as a geologist the equal of his elders.

And Yount paid him accordingly: two years after Halbouty went to work for Yount-Lee, the company was paying him more than $600 a month. A year later he was making $750, a salary even divisional geologists for some major oil companies could envy.

Yount guided Halbouty in almost every aspect of his life except the social, and in that area Halbouty required no help. He was a rounder, and there were few towns or villages near the company's widespread holdings where he could not find a pair of welcoming female arms. Young women in Beaumont who knew him whispered about him and called him a fast worker.

In the summer of 1933, at a neighborhood party in Beaumont, he met a newcomer from Fort Worth, Lesly Carlton. She was Halbouty's age, a woman of astonishing comeliness with dark hair and brilliant blue eyes, a combination to inspire poets and stir a young man's blood. She had come to Beaumont to visit a married sister; she stayed on because of Halbouty, though it was two years before she told him this.

Halbouty saw Lesly Carlton as often as he could when he was in Beaumont, but he still kept warm his relations with women he knew in High Island, Barber's Hill, Sour Lake, and several Louisiana towns. The world was his oyster, but the oil business was his pearl.

Halbouty was working in the Jennings, Louisiana, field in November of 1933 when he was ordered to return to Beaumont immediately; Miles Frank Yount had died suddenly of a heart attack in his fifty-third year. Halbouty couldn't accept the message at first. Yount had appeared to be in vigorous health, and Halbouty had never heard him complain of an ailment, major or minor. But the truth broke through as he drove toward Beaumont. He had never suffered the loss of a relative or a close friend, so grief of that nature was unknown to him. Now he felt as if his heart would break. He parked his car on a side road and cried the first tears he had shed as an adult.

If Yount's death shocked Halbouty, it was a demoralizing blow to the city of Beaumont. Yount had been a kind man, and his generosity had been legendary. The past Christmas of 1932 had found the city unable to meet its payroll. Yount had met it so that city workers could have a good holiday. But to the city's leaders, Yount's death meant the end to a dream of economic security. They knew that Yount already had embarked on a program to make his company a major integrated oil company, and they had been assured that he would not move it to another city. Work already had begun on the largest oil-tank farm in the world. Yount had acquired reserves of oil as a backlog; he had started a pipeline network and a deep-water terminal. But Yount was the heart and soul of his company; without him, the city leaders believed, the great dream would disintegrate.

Yount had run the company with little help and no interference from the handful of people who shared ownership with him. They ran the company as best they could after his death, moving forward on the momentum Yount had generated, but the time came when they were ready to sell. Several major companies made tentative proposals, but a deal finally was made with Standard Oil of Indiana through its drilling and producing subsidiary, Stanolind Oil and Gas Company.

Both sides agreed to rely on an outside appraiser to set a value on the Yount-Lee holdings, and a nationally-known geologist, D'Arcy Cashin, was selected for the job. W. E. Lee, one of the Yount-Lee partners, offered Halbouty's services, but Cashin declined. "I think I'll be able to find everything," he said.

Lee was a big, robust man in his fifties who lived in Sour Lake, Texas, and commuted to Beaumont as Yount-Lee business demanded. He had grown closer to Halbouty during the long months after Yount's death, and depended on the younger man to help him make decisions. Now he asked Halbouty to make a private appraisal of the Yount-Lee holdings. "Just to give me an idea, Mike."

Cashin's official evaluation was $48 million.

Halbouty's private evaluation, told only to Lee, was $135 million. Lee was flabbergasted at the difference.

Officials of both companies met in the Yount-Lee boardroom to hear Cashin's evaluation and complete the deal. Lee sat across the long table from F. O. Prior, Stanolind's president. Cashin sat next to Prior and Lee asked for Halbouty to sit next to him. The other officers were seated on down the table.

Cashin delivered his evaluation. Lee spoke up. He told the gathering of Halbouty's evaluation. "I'm amazed at the difference," Lee said, "and I wonder if it wouldn't be proper to have Mr. Cashin explain in more detail how he arrived at his figure."

The remarks enraged Prior, the Stanolind president. "He's *your* geologist," he said, pointing at Halbouty. "If we had used *our* geologist, he might have come up with a figure of twenty-four million. But we've used an impartial geologist, as we agreed to do, and he's come up with his figure."

Lee turned to Halbouty. "What do you have to say, Mike?"

"I know the properties, and Mr. Cashin doesn't," Halbouty said. "For instance, we drilled a well on our Hackberry Island, Louisiana, property and found two hundred feet of oil sand. But the hole caved in, and we didn't drill again because Mr. Yount wasn't here to make us do it. But the oil is still there. Mr. Cashin is ignoring the oil sand and valuing the property at its lease worth. Mr. Cashin has been using my maps, but he didn't use all my log records, and he didn't read all my reports on corings. He just doesn't know as much about the properties as I do."

Halbouty's last sentence set a fire along the table. Prior shouted at Lee, other officials shouted at counterparts across the table, and Cashin, in all of his gray-haired dignity, stood up and shook his finger in Halbouty's face. "Young man," he shouted above the hubbub, "how many evaluations have you made?"

"Just this one," Halbouty said.

"I've made thousands! Thousands! So keep your mouth shut!"

"Yes," Prior said, "and while we're at it, I don't think we need him in this room!"

Lee nodded at Halbouty and the young man left the office. Inside, the bullyragging continued, with Lee proposing another evaluation by another geologist. Prior refused to listen. And finally the deal was made, for $48 million, simply because the Yount-Lee partners wanted out.

As the years passed it became evident that even Halbouty's evaluation was extremely low. The old Yount-Lee properties produced billions of dollars worth of oil. The subject became W. E. Lee's favorite topic of conversation. "It was the greatest giveaway in history," he would lament to his cronies. "We gave away billions!"

The deal was not consummated for weeks, and in that period Halbouty was summoned by the Stanolind Personnel Chief. "You're going to have to take a pay cut if you stay with us. And we're going to have to do something about your work category."

"What do you mean?" Halbouty asked.

"Well, there's nothing in our table of organization for a man who is a geologist *and* a petroleum engineer. You're going to have to be one or the other."

"I'm not going to throw away one of my disciplines just to fit into some damned silly table of organization," Halbouty said. "You can take your job and stick it up your ass. I'll be leaving when Stanolind takes over."

He received job offers from several major companies, but he turned them down. Then W. E. Lee came to see him. Lee's son-in-law, Glenn McCarthy, had gone into the oil business several years earlier, and Lee wanted to help him. He recently had loaned Halbouty to McCarthy to work the paleontology on a well McCarthy had drilled near Anahuac, Texas. The well had come in strong just a week before. Now Lee wanted Halbouty to go to work for McCarthy.

"He's mentioned it to me, Mr. Lee," Halbouty said.

"Well, he can't pay you but three hundred a month, but

I'll make up the difference so you'll get what you were getting with Yount-Lee."

"Nobody's going to pay me under the table, Mr. Lee. You have Glenn come and talk to me. If we can figure out something where I've got an interest, the salary won't be too important."

So in late July of 1935—four years after he had signed on the Yount-Lee payroll—Halbouty met with Glenn McCarthy and they decided to work together out of Houston.

It was something like King Kong and Godzilla agreeing to share the same apartment.

three

Like Halbouty, Glenn McCarthy hailed from Beaumont. Like Halbouty, he had worked in the Spindletop field as a waterboy. His father was an oilfield hand, and McCarthy grew up around the various fields his father labored in. He was working in a Houston service station in 1930 when he eloped with W. E. Lee's teenage daughter, Faustine. It was another two years before he went into the oil business, and then it was on a shoestring. The shoestring was a battered, weary drilling rig with other tools and equipment he could borrow or his crew could borrow by moonlight. He had the guts of a pit bulldog and strength of body no one would believe until they saw him use it. His temper was dangerously uncertain, but he could charm Lady Godiva off her horse or a pool hustler out of his last quarter.

McCarthy was already well known around Houston and in the Gulf Coast oil patches. He was a larger man than Halbouty, in both height and weight, but built on the same trim lines. A lock of dark hair fell across brooding eyes in a rugged face. He was known as a man who could drill a well in half the time it would take a major oil company; a man inclined to raise his fists at every affront whether large, small, or imaginary.

He was twenty-nine to Halbouty's twenty-six when Halbouty joined him in his office in the Sterling building at Texas and Fannin in downtown Houston. McCarthy had a little oil production in the Conroe, Big Creek, and Anahuac fields, but most of his income went to pay old debts and meet current expenses when he could. And he had a couple of venerable drilling rigs.

But he was an oilman, possibly the best practical oilman the country had produced, an improviser who could drill with junk, devising equipment on the spot to replace broken or worn-out standard gear. He had a knack for finding oil he couldn't explain because it came installed in his system like an antenna. And he was a plunger, always willing to shoot the moon on his chances of finding oil. All these things Halbouty admired about him, and for a time they outweighed what he considered defects in McCarthy's character. Their deal called for Halbouty to receive three hundred dollars a month and a percentage of what they found after costs of finding and producing were paid.

They began their first venture almost immediately. McCarthy said they needed oil. Halbouty unfolded a map and pointed to an area west of Beaumont. "There's oil. I know that area. I studied it when I was with Yount-Lee, and a geophysical crew found a structure there."

McCarthy pointed out that the Amelia oilfield was only a few miles from the area, and that it was common knowledge that the Amelia oil sand petered out against a big fault in the direction of the land Halbouty was touting.

"True," Halbouty said, "but Amelia's on the down-throw side of the fault. We'll be on the up-throw side."

McCarthy told him to check it out. Halbouty walked to the Gulf Building on Main Street and went to the office of August Selig, Stanolind's divisional geologist. The men had become friends during the long months of negotiation between Stanolind and Yount-Lee.

Selig greeted Halbouty warmly. "It's a funny thing, Mike. I was just getting ready to call you. We've got a hundred-acre lease west of Beaumont, and it's about to expire. I thought maybe you and Glenn might want to drill it. All of our rigs are busy."

Halbouty laughed. "I came over here to talk to you about the same area. When does your lease expire?"

Selig grunted. "In twelve days if we don't drill."

"What!"

"Twelve days."

"Goddamn! Are you serious?"

"I sure am."

McCarthy's two drilling rigs also were busy, drilling for others on a contract basis. But Halbouty asked, "What's the deal?"

"Oh, we'd just take an override," Selig said. This meant that Stanolind would receive one-sixteenth of any oil produced from the lease. The land owner would receive two-sixteenths and McCarthy would receive the remaining thirteen-sixteenths. McCarthy would have to stand all of the expenses of drilling, whether or not oil was found. "If you don't take the deal, Mike, I'm going to have to start talking to somebody else right away," Selig said.

"I've got to talk to Glenn, but go ahead and put us down for it. I think I can get him to take it."

"Suppose he doesn't want it?"

"Then I'll have to come back and tell you. Just don't do a damn thing with it!"

Selig agreed to wait, and Halbouty returned to the McCarthy office. He told McCarthy about the deal—and told him he had taken it. McCarthy exploded. He cursed and ranted. Halbouty cursed and ranted. (A visitor asked the telephone operator, Maxine Manuel, what the hell was going on. "Oh," she said lightly, "Mr. McCarthy and Mr. Halbouty are just having their regular daily conference.") Finally Halbouty outshouted McCarthy long enough to say, "If you don't want the deal, just say so—and I'll go back and tell Selig the hell with it!"

McCarthy sat down. "Let me think about it," he commanded. He had a lot to think about. Drilling a rank wildcat on such a small tract of land was almost unheard of; it is customary to lease all available acreage around a prospect. It was raining all along the Gulf Coast; moving in and setting up a drilling rig in eleven days in gumbo mud would be impossible. And he didn't have a rig available, anyway. So he called George O'Leary, president of Houston Oil Field Material Company, and shortly O'Leary arrived at the office.

Halbouty watched with admiration while McCarthy convinced O'Leary that he should sell McCarthy a seventy-five-thousand-dollar drilling rig, ship it to a location west of Beaumont, and assemble it in eleven days . . . all on credit! (In later years O'Leary would write that McCarthy ". . . was a consummate trader willing to stand or fall on his ability to match wits with all comers.")

O'Leary called his office and got work started. And it kept on raining, the kind of downpour Texans call a "frog strangler." Huge trucks bogged down in the mire on the lease; mules and men inched them toward the spot Halbouty had selected as the drill site. McCarthy's crew and O'Leary's people worked shoulder to shoulder, and no one worked harder than McCarthy and Halbouty. They worked around the clock, day after day, and still it rained. The rain blinded them as they handled the slippery steel to fashion the derrick.

They had a visitor, L. P. Teas, chief geologist for Humble.

His company had brought in the nearby Amelia field. "I hope we find a field just like Amelia," Halbouty told him.

"Ah, Mike, lightning doesn't strike twice in the same place," Teas said.

"It's a different lightning," Halbouty said.

Teas shrugged. "You're not going to hit this time, Mike."

When Teas left, McCarthy said to Halbouty, "You son of a bitch, we'd *better* hit something. I've gone in debt up to my ass just because you said the oil is there."

Louis Baker, McCarthy's drilling superintendent, shook his head woefully. "We may not get a chance to find out if there's oil here. I don't think we'll beat the deadline."

"We'll beat it," McCarthy said. "Some way."

On their last day of grace they couldn't get the engine in place to attach it to the rotary table that turned the drill pipe. Hours passed. The lease expired at midnight, and in the late evening McCarthy arranged for some notaries public and city officials to come to the site. He still intended to beat the deadline, and he wanted it officially witnessed.

At eleven o'clock that night the task looked hopeless. The engine was still not attached to the rotary table. McCarthy looked around at the weary, mud-crusted men. "All right," he said softly, "we're going to start drilling this son of a bitch by hand."

It was still raining when, at fifteen minutes before midnight, they finally sunk the bit into the earth, turning the drill pipe by hand with chain tongs. City officials signed affidavits, duly notarized, that Glenn McCarthy had beaten the deadline.

The next day the sun came out. A man in a black suit with a black wooden box in his hands slogged through the mire to Halbouty's side. "You want me to tell you whether you're going to make a well or not?" he asked.

"Yeah," said Halbouty, eyeing the box.

The man went to the rig floor and ducked under it. In a

few minutes he slogged back to Halbouty. "I hate to tell you this," he said, "but you're going to make a dry hole."

"Are you sure?"

"Of course I'm sure!"

Halbouty cupped his hands to his mouth and shouted at Louis Baker who was working on the rig floor with about a dozen men. "Louis! You guys stop working!" All work ceased, and all hands turned to face Halbouty. He shouted, "This doodlebugger says we're going to make a dry hole, so start tearing down that damned rig!"

It took a long moment for Baker and the others to grasp the point; they had been working harder than slaves to get the well underway, and now they were being ordered to abandon it. Then Baker started laughing and the others joined in; a wave of merriment rolled over the prairie, sweeping the doodlebugger along with it. "Fools!" he muttered with disgust as he trudged away.

Despite the doodlebugger's prediction, the well came in flowing 288 barrels of good oil per day through a 1½-inch choke. It was the Longe 1, the discovery well of the West Beaumont field. Stanolind was happy, too. The company owned other leases all over the general area.

Halbouty had been spending what free time he could muster with Lesly Carlton in Beaumont. In mid-October, after he had been with McCarthy eleven weeks, he startled her by asking, "Why don't you come to Houston?"

She stared at him. "What?"

"Oh, hell," Halbouty said. "Will you marry me?"

Lesly started laughing.

"What's so damned funny?" Halbouty demanded.

"Ah, Mike," she said, "I knew if you asked me, it would be something like this."

"Well, will you do it?"

She nodded. "You know I will."

The newlyweds moved into Halbouty's upstairs apartment on Fannin Street. It seemed almost immediately that Lesly was struck by arthritis that crippled her fingers. Her religious beliefs precluded medication or medical treatment. Halbouty pleaded with her. He stomped the floor. He called on her relatives in Fort Worth for help in persuading Lesly to see a doctor or enter a hospital. He received none. The disease moved quickly to other parts of her body, and she could not easily negotiate the stairs to their living quarters. She remained serene in her faith.

Halbouty was working from can to can't, both he and McCarthy hunting new prospects while drilling the leases McCarthy held, but he needed money to make life easier for Lesly. He had found a small stucco house on Hawthorne Street in the Montrose section of the city whose owner agreed to sell for six thousand dollars with a fifteen-hundred-dollar down payment. Halbouty had a thousand dollars, but he had little hope of saving the needed five hundred dollars on his three-hundred-dollar-a-month salary.

So he went to see Warren Baker, brilliant editor of *The Oil Weekly* (now *World Oil*), with a proposal that Halbouty write a series of articles for the magazine that would then be incorporated in a book. Baker liked the idea. He had bumped into Halbouty at several oil gatherings, and he had heard him praised by major-company geologists. He took Halbouty in to see his publisher, Ray L. Dudley, president of the Gulf Publishing Company. Halbouty was eloquent; Dudley was impressed. Halbouty said he would write seven articles for the magazine, then work them into a book—for a five-hundred-dollar advance on royalties. Dudley wrote him a check and the next day Halbouty bought the house on Hawthorne Street. There were no stairs to confront Lesly.

The book had a formidable title: *Petrographic and Physical Characteristics of Sands from Seven Gulf Coast Producing Horizons.*

The research was fresh and the writer confident though his prose wobbled.

> The primary purpose of this study [Halbouty wrote in a preface] was to determine whether or not certain geological zones were correlated by this method, and this purpose was accomplished to some extent to a surprising degree of accuracy. It is also well to state in this preface that the secondary purpose of this paper could easily be one of creating interest in this kind of study which would, after much research and in later years, prove to be profitable to the oil geologist in his endless search for oil-bearing structures.

Scientists were not baffled by his sentence structure, and the book had a brisk sale. Like Halbouty, they believed that

> stratigraphical traps exist in the Gulf Coast which show no relative "high" by the use of a geophysical instrument, and which will show no feature by correlative methods based upon paleontological guides. Therefore, it seems very plausible, as well as logical, that the only means that the geologist can turn to for information that may lead to criteria to form a stratigraphic trap would be by the employment of sedimentary and stratigraphic factors. These sedimentary and stratigraphic factors can be obtained only by the study of the sediments, and especially by the study of the sands . . .

While the new drilling rig was being skidded to drill Longe 2, one of the old McCarthy drilling rigs was placed on a lease near the village of Winnie, between Beaumont and High Island. Halbouty had studied the area—it was rank wildcat territory—and found what he considered a favorable locality for the accumulation of oil and gas.

When the well was down to 6,000 feet, the driller started out of the hole to change bits. Halbouty was on hand to take samples of the deep drilling. He was standing a short way off when he heard a thundering noise followed by the shriek of twisted steel. As he watched, the entire derrick collapsed. The derrick man, working on the monkey board high in the air, was crushed to death before his eyes. The other crewmen had darted for safety.

Halbouty's years in the oilfield had made him aware of violent death, but he was never able to accept it. Here, on one of his own wells, he felt an almost personal responsibility for it. Yet he realized it was an accident over which he had no control. The driller had started to lift the drill pipe out of the hole without unlocking the blowout preventer. The drill pipe couldn't budge, so the derrick had to give.

The loss of the derrick at Winnie temporarily halted drilling in that area. And that loss was quickly followed by another. While Halbouty and McCarthy were resting in a Beaumont hotel, Halbouty gazed out a window to see a bright reflection in the western sky. He roused McCarthy. "Glenn, that looks like a fire near where we're drilling!"

It was. Longe 2 had blown out and was on fire. By the time they reached the well site, the new derrick was melting in the flames. The crew was trying, without success, to move the major equipment away from the fire. No one had been hurt.

McCarthy called in Myron Kinley, the famed "hell fighter" whose protege, "Red" Adair, would win an international reputation in the hazardous business in later years. Kinley and his intrepid crew managed to save the major equipment, and another derrick was brought in immediately. The equipment was joined with it, and McCarthy began drilling another well even before the wild well was brought under control.

The wild well brought on a number of lawsuits. It was claimed that the fire and its fall out had damaged homes, and that chickens refused to lay, cows to give milk, fruit and nuts to grow. The main case, the one tried in court, contended that land near the well had been rendered unproductive for all time by spraying oil and gas.

McCarthy was defended by a brilliant young Houston lawyer, Leon Jaworski. For his purposes, Jaworski obtained several postponements in the trial. And then McCarthy was injured and had to undergo surgery. The delays allowed Jaworski to work his magic; he had been holding out for growing weather. To prove the land was arable, he showed color pictures of trees in leaf and tall green grass. He even had a picture of cows munching contentedly at the flora. It marked the first time such evidence was introduced in a Texas court.

His case was endangered, however, when he summoned McCarthy to testify. Jaworski had been describing McCarthy as a young man in dire financial straits. McCarthy made his courtroom appearance in a wheelchair pushed by a well-dressed black man—"His valet!" the onlookers whispered—and accompanied by a beautiful blonde in a nurse's uniform. McCarthy himself was attired in a silk robe with a silk scarf at his throat. It took all of Jaworski's skill to win the verdict for the wildcatter. Most of the other cases were dropped, though a few were settled for small amounts.

During the long months between the blowout and the end of the trial, Halbouty led the way to Cotton Lake, still in the general Beaumont area, and McCarthy's drillers brought in Kilgore 1, the discovery well of the Cotton Lake field. After he had drilled eight wells on his lease, McCarthy sold out for a net profit of $750,000 to help pay debts and gain some working capital. He also sold his Anahuac holdings for a net profit of $500,000, but continued to drill in the field on a contract basis for the new owner.

Though McCarthy had drilled at Anahuac before Halbouty joined him, Halbouty sat in on the sale negotiations. When the session was completed, and the two men were in a corridor, McCarthy growled, "Goddamnit, you almost kicked my leg off in there. It's going to be black and blue."

"You gave it away," Halbouty said.

"I needed the money, damn it."

"You could have borrowed that much at a bank."

"Let's go," McCarthy said.

During this time Halbouty also underwent surgery, an emergency appendectomy in Beaumont. The surgeon who removed the swollen organ left a sponge in its place. Halbouty was in critical condition when another surgeon decided to re-open the incision and found the sponge. While Halbouty was recuperating, McCarthy reorganized the company. He visited Halbouty in the hospital, explained what he was doing, and Halbouty signed papers waiving his percentage in lieu of stock.

And during this time, as they had from the beginning of their association, the two men continued to quarrel over almost every aspect of their activities. It was said jokingly in oil circles that when they held a discussion in the Sterling Building they could be heard clearly at the far corner of the Rice Hotel at Main and Texas, a block away. Many of the arguments were in Halbouty's laboratory, where he studied corings and cuttings from the various wells. The two men, it seemed, were in constant dispute over composition of the formation material. "They'd fuss until Mike got his books and showed Glenn what he was talking about," the telephone operator, Maxine Manuel, told an inquirer. The "cuss fightin'," the always being at odds, did not appear to overly disturb either man. But Halbouty was getting fed up with what he considered McCarthy's excesses—actions and attitudes that others in that milieu and time may not have judged so severely.

Soon the two men moved into the area around Palacios, a small town some one hundred miles down the coast from Houston in Matagorda County. This, too, was wildcat territory. The first well, while it was still being drilled, convinced both men they were boring into one of the great oilfields of the country. The bit found a familiar gas sand thirteen hundred feet higher than it was normally located on the coast. With other favorable structural conditions, this promised that the wildcatters were near the Frio sand, which was rich with oil in some other coastal fields.

Oil companies and wildcatters alike flocked to Palacios on the strength of the gas showing, buying leases with cash. McCarthy tried to stay ahead of them, even leasing town lots in an effort to build up a big block of acreage. He spent more than seven hundred thousand dollars on leases, and went a million dollars in debt for five drilling rigs. Then, consummate trader that he was, he sold a half-interest in his leases for five hundred thousand dollars to a major oil company to get drilling money.

The five new rigs were put to work and the original well, the Foley 1, was sunk to where the Frio sand should have been productive . . . but it wasn't.

The bits at the other five wells reached the gas sand at about the same time, and each well threatened to blow out. It took tremendous loads of synthetic mud to keep them under control, forcing McCarthy deeper into debt. All were finally completed as gas wells, in an area where there was no market for gas, and at a time when the output of a gas well would hardly pay for drilling costs.

McCarthy was left with an accumulated debt of one and a half million dollars, and his several hundred creditors were crying for his shirt. He probably would have been driven to his knees had it not been for the Longe lease at the West Beaumont field. Sixteen wells, all good ones, had been drilled on the lease after the fire at Longe 2. McCarthy put up the production from the lease as security for his indebtedness, and put all of his rigs into contract work. He was through with wildcatting—for a while.

Meanwhile, Halbouty had parted company with McCarthy. McCarthy implied that Halbouty had left him because he thought McCarthy would never recover from the Palacios debacle. But Halbouty certainly knew the worth of the Longe lease, and he had unbounded confidence in McCarthy's ability as an oilman. He said he simply had seen too much of what he considered the darker side of McCarthy's nature.

It was one of the world's wonders that the two men did not come to blows themselves during their nineteen-month association. Indeed, Halbouty on occasion played the peacemaker in conflicts between McCarthy and others.

McCarthy recovered from Palacios. He returned to Winnie, where the collapsing derrick had killed a crewman, and found great quantities of oil and gas. He discovered other oilfields, built plants and factories, plunged into businesses other than oil, and crowned his achievements with construction of Houston's Shamrock Hotel.

For a while he was the most publicized oilman in the country, a darling of the news media as much for his flamboyant lifestyle as for his business activities. He participated in civic affairs, contributed to and worked for charities, focusing national attention on Houston with almost every public word and act. He was on the cover of *Time* Magazine.

But in the late 1950's, his empire began to disintegrate under pressure from creditors to whom he owed more than fifty million dollars. Bit by bit his possessions were taken from him, even to the Shamrock, his pride. There was one last try for the big time—a widely publicized wildcatting expedition into Bolivia—and it was a disaster.

"They would have been a hell of a team if they could have stayed together," oilmen said when news of Halbouty's leaving McCarthy made the rounds. "They could have been as big as they wanted."

Halbouty didn't go far when he left. He moved across Fannin Street into the Shell Building—and began for the first time to learn about dry holes and taste the bitterness of failure.

four

HALBOUTY was on his own, and he was broke. He took nothing with him when he left McCarthy, neither percentage, stock, nor cash settlement. He rented a single office and partitioned it into three segments with room dividers—a place for a secretary, a place for himself, and a place for the landman he hoped to hire.

He had two things in his favor—a reputation for finding oil wherever he sought it and the self-confidence of a boxer who has never known defeat. Both would suffer greatly, but not before he had enjoyed a brief triumph.

It began with a phone call from an old friend, B. L. Woolley, who was in town from Dallas and wanted lunch. Halbouty was a junior at Texas A&M when he met Woolley. Halbouty had been elected to edit his class annual. Woolley was then a salesman for a Missouri printing firm; he bid on the printing of the annual and won the contract. He was a colorful man who liked a nip of bourbon and Havana cigars. He and Halbouty had become friends.

At lunch they brought each other up to date. Woolley had long since left the printing business and had become a millionaire through investments. His brother, B. W. Woolley, also had become wealthy in the same manner. Over coffee, Woolley said he and his brother would like to invest in Mike Halbouty.

Three weeks later the Merit Oil Company was formed. The Woolley brothers held a one-third interest; Halbouty held the remaining two-thirds. He gave a piece of his interest to Johnny Chance, a drilling superintendent with whom he had worked at Yount-Lee, and with another piece he attracted a former McCarthy landman, Alec Dearborn.

Halbouty knew that Humble geophysical crews had been quietly "shooting" a large area near Friendswood, a community east of Houston. As secretly as possible, he conducted his own geological survey. Then he put Dearborn to work checking land ownership. Dearborn quickly learned that Humble had leased up all of the area save for several small tracts.

Halbouty called in a rotund, jovial, back-slapping "lease hound" named Waldo Cahill. Cahill went "poor-mouthing" among the tract owners, exuding his particular brand of wit and geniality, but he could lease only one of the plots—and it held only thirty-seven acres. "Get it," Halbouty told him. "It'll cost fifteen hundred dollars," Cahill said. "Get it," Halbouty repeated.

He wanted more acreage before he drilled anywhere, so he began surveying another area within cannon shot of Houston, near a settlement called Cedar Bayou. When Humble learned that Halbouty had obtained the lease at Friendswood, it sent a man to see him. The Humble geologists, who knew Halbouty's track record, were convinced he would be hard to deal with. "He doesn't think he can drill a dry hole," one reported to the Humble brass. "He intends to drill that lease and drain our oil. He won't sell."

For weeks, while Halbouty continued to study the Cedar

Bayou area, Humble continued to dicker with him. Finally the company offered him $125,000 cash plus $75,000 in oil produced from the lease. He snapped up the offer. The partners in Merit Oil Company were overjoyed at the *coup.* "What a beginning!" B. L. Woolley wrote Halbouty. (Humble later brought in a fine field at Friendswood—but the thirty-seven-acre tract was barren.)

With money in hand and support from the Woolley brothers, Halbouty went out after other promising leases. He splurged, hiring a geophysical crew to "shoot" some areas. And he took on a new employee, a petroleum engineer fresh out of Texas A&M, Milton Cooke. Cooke had worked for McCarthy, under Halbouty's direction, during summer recesses from school.

Halbouty drilled eight dry holes in a row, in four different areas. One of the areas was Spindletop. He was drawn back there by the mound's mystique; he wanted to succeed where the pioneers had introduced the Liquid Fuel Age, and where his idol, Miles Frank Yount, had been named the "Master of the Mound." His three wells found not oil but salt water.

At Crosby, Cedar Bayou, and Hitchcock, all near Houston, he found shows of oil. Major and larger independent companies, more patient and better prepared, moved in after him and discovered good fields at all three locations. In some instances they followed because they had confidence in his ability to find oil but considered him in too much of a hurry to take proper advantage of his talent.

At Cedar Bayou, while his record was still unblemished, he came into contact with a remarkable woman. As he studied the area and went about leasing acreage, he found that some desirable land already was held in lease and in fee by Mellie Esperson, widow of a Danish wildcatter, Niels Esperson, who had made a fortune in oil soon after the Spindletop boom in the early 1900s.

Esperson's first strike had been at Humble, north of Houston, where he had drilled three dry holes on a five-acre lease before bringing in a gusher. Mellie had worked in the field with him, intimately sharing the heartbreak of the dry holes. After his death in 1922, she had carried on his various enterprises, increased the family fortune substantially, and more than held her own in the materialistic society of a booming Houston. A newspaper business editor described her as "formidable."

Halbouty found her gracious. She was in her late sixties, a sturdy woman with snow-white hair brushed straight back from a high forehead. She had offices in the Niels Esperson Building, a thirty-two-story skyscraper. It had been the tallest building in Texas at its completion in 1927. (She would build a nineteen-story building adjoining it, and name it for herself. "Not as tall as my husband's building," she instructed the architect.)

Halbouty unrolled his map on her desk. He showed her his leases abutting her acreage. He explained his geological conclusions. She looked at him sharply. "You haven't seismographed your leases?" she asked.

"No, ma'am."

She didn't ask him why. Instead, she began asking him personal questions—questions about his parents, his boyhood, his schooling, his wife, his short years in the oil business. When she was quite through, she said, "When my husband started in the oil business, banks didn't lend money to wildcatters. But he finally found a banker who backed him. The banker didn't know anything about the oil business. He loaned the money to the man, my husband."

"Yes, ma'am."

"Now, you go see my geologist. He's Alexander Deussen, and he has an office in your building. I'll call him and tell him to work out a deal with you." She put out her hand for

Halbouty to shake. "You've had a good life, so far, Mr. Halbouty. I hope the rest of it is."

Deussen was a respected geologist in his sixties. He had read Halbouty's book and several articles the younger man had written for trade magazines. When he examined Halbouty's map, he did not question his geological deductions. He slapped a hand on the map. "Let's drill a well, Mike!"

They made a deal to drill at least three wells on their pooled acreage, sharing the expenses and the profits, if any. An Oklahoma drilling contractor was brought in to do the work. At six thousand feet the drillers found the Frio sand, and it was bearing oil. Halbouty was on hand to help take the core. B. L. Woolley was in Dallas, but his brother, B. W., was on the rig floor. B. W. was as phlegmatic as his brother was ebullient. "You could bust a firecracker in his ass and he wouldn't even blink his eyes," a roughneck said of him.

Halbouty picked up some of the core in both hands and held it out for B. W. to look at. "See that," Halbouty crooned. "Isn't that a lovely sight?"

B. W. grunted.

Halbouty lifted the core and washed it across B. W.'s face. "Smell it, you sourpuss son of a bitch!" The crew roared with laughter. B. W. got out his handkerchief and mopped his face. "It's nice, Mike," he said solemnly, and again the roughnecks roared.

It was nice, but Halbouty wanted more. He instructed the drillers to take the well down to seventy-five hundred feet. The Vicksburg sand normally was found near that depth, and he hoped to find it saturated with oil. The drillers found the Vicksburg, but it was dry.

It was time to produce oil from the Frio at six thousand feet. But a well-logging company mismeasured the well by fifteen feet and perforated the casing not in the Frio but in a salt-water formation below it. The well began producing salt water. The well was "squeezed"—cement was pumped into the well so that it would rise between the casing and the earth walls and seal off the water from the perforations. The squeezing failed, and so did months of squeezing that followed. And the squeezing finally rendered the Frio formation incapable of production. The well was capped. There was talk of suing the well-logging company, but no action was taken.

Meantime, Halbouty was getting dry holes with monotonous regularity. The time came when the money was gone and the Woolley brothers' tolerance had run out. Halbouty and Milton Cooke met with the brothers in the Texas State Hotel near Halbouty's office. Halbouty was now twenty-eight. His pride was aching, and he spoke out of his bitterness. "You guys haven't got the guts for this business," he told the brothers. Said B. L. Woolley, his friend of years, "We're not in as big a hurry as you are, Mike."

That was the end of the Merit Oil Company. Halbouty left them and drove to Cedar Bayou where the drilling rig on the third well—a dry hole—was being dismantled. He had six dollars in his pocket, all the money he possessed. He drove to nearby Baytown and bought a bottle of Johnny Walker Scotch. He drove back to the well site. He sat there in his car, drinking the whiskey, watching the derrick come down. He was not a heavy drinker, but he drained the bottle without getting even a buzz. When the derrick was down, he began the drive to Houston and home.

It was not his normal nature to blame himself; his ego would not permit it. But tonight he could not escape the truth that he had allowed his arrogance and impatience to betray his knowledge and his judgment. He told a friend: "I had forgotten everything Miles Frank Yount had taught me. He had told me never to fear going in where others feared to tread, but to be sure I knew what I was going to do once I got there. He told me never to take shortcuts, and I had taken them by the dozen." Most painful to him was the knowledge that by taking shortcuts he had left in the ground oil that should have been his. (At a spot less than one hundred feet from his last dry hole at Cedar Bayou, the Commercial Petroleum & Transport Corporation would bring in a well that would be the official discovery well of the field.)

Lesly was waiting up for him. She was in a wheelchair now. They drank coffee, and she read the defeat written on his face before he poured out the bitter story. He was concerned about Mellie Esperson. "She thought I was wasting money squeezing the first Cedar Bayou well," Halbouty said. "She probably thinks I'm a fool."

"Perhaps you have been," Lesly said.

Halbouty stood up from the table. He began pacing the floor. Lesly sat silently; she thought she had triggered his temper and that he was fighting to control it. But suddenly Halbouty stopped his pacing and looked down at her. "Lesly," he said, "this is the best thing that could have happened to me."

She raised her brows.

"Look," he said, "it's the best thing. I've been cocky. I've been going around with my nose in the air. I've been telling myself that I was going to be the biggest thing in this business. I've been taking it for granted that everything I touched would work out. Now I know it won't. I'm not sure about everything anymore, but I'm sure about one thing. If I had brought in an oilfield on the first time out on my own, I would have thought I was so great I wouldn't have been fit to live with. It would have destroyed me."

Lesly said, "God's been chastening you, Mike."

"I guess."

"What are you going to do?"

Halbouty grinned. "I'm going to start using my brains for a change."

The Merit Oil Company name had come off the door and in its place was the legend MICHEL T. HALBOUTY, CONSULTING GEOLOGIST & PETROLEUM ENGINEER. Alec Dearborn was dead. Milton Cooke was on his way to Venezuela to drill directional wells for Gulf. Johnny Chance had found a job in Louisiana. Only the secretary, Pearl Mountz, remained on the payroll, and she had taken a pay cut from $125 a month to $85. The Consulting Geologist & Petroleum Engineer was meeting his payroll and feeding Lesly and himself by writing articles for trade magazines. Most of the articles he wrote by himself; others he wrote in collaboration with recognized older scientists who were happy to have his help.

Pearl Mountz almost ran off his first customer. "There's a man out there in my office in dirty overalls and with manure on his boots. Do you want me to get rid of him?"

Halbouty snorted. "Hell, no! I haven't talked to anybody but you for two weeks. Show him in."

The visitor was chunky and unshaven. He said he was Charley Alexander, that he was a dairy farmer, and that he owned some acreage near the Hastings oilfield about twenty-five miles south of Houston. He unfurled a map and showed Halbouty his acreage with a rusty forefinger. "I've been to every oil company in town, but nobody wants to drill my land," he said. "I read your write-up of the Hastings field in one of those oil magazines, and it seemed like you thought there might be oil in the direction of my land."

Alexander was right. Hastings was a giant field, but Halbouty believed its limits had not yet been reached. And Alexander's land lay in one of the directions he believed the field could be extended beyond existing production. "I thought maybe you might want to drill it," Alexander said. "Or maybe you can get somebody else to drill it, with your reputation and all."

Oh, how Halbouty wanted to drill it! And he was also interested in the man. Alexander was a farmer, but he wasn't a bumpkin. He spoke with the quiet certitude of a man capable of taking care of his interests. "Give me a ninety-day option and I'll see what I can do," Halbouty said.

"Thirty days," said Alexander.

"Make it sixty."

"Forty-five, and that's my limit."

"Good enough," Halbouty said.

They shook hands on the deal.

Halbouty scooted around Houston like a waterbug on a fresh pond. He went to wildcatters, pure promoters and all the oil companies, independents and majors alike. He could generate no interest—and his option expired. A short time later Stanolind Oil & Gas dealt with Alexander, and the wells drilled on his acreage made him a millionaire.

He came by to see Halbouty. He was clean-shaven and his boots were polished, but he still wore overalls . . . this pair an off-white. "I wore overalls all my life, before I got rich," he told Halbouty, "and I guess I'll wear them from now on." He had come by, he said, to thank Halbouty. "Nobody ever would have drilled my land, if you hadn't got everybody interested."

(Alexander *did* wear overalls all of his life. Alva W. Bounds, a Stanolind attorney who dealt with Alexander for the company, took Alexander on trips to Chicago, St. Louis, and other cities. Alexander liked to ride on trains, and he wore overalls even on these trips. "He doesn't carry a thing with him," Bounds told a friend. "When he wants something—a shave, a clean shirt or overalls—he just buys it.")

Losing the Alexander option was a terrible blow to Halbouty. "God's still testing you, Mike," Lesly said.

If so, God also was testing Halbouty's new friend, a young lawyer, Harry Sims, who had an office across the street in the Sterling Building. Sims was trying to specialize in oil law, but he had as few clients as Halbouty. Still, they talked big about the future over fifteen-cent bowls of chili at lunch. One day Sims came to Halbouty's office with a question about some acreage west of Houston. Halbouty dusted off his maps as if Sims was an important client. He gave his opinion of the acreage in question. When Sims left, Halbouty told Pearl to send Sims a bill for $125.

Sims seethed, but said nothing about the bill. A few days later Halbouty called Sims to inquire about a paragraph in a lease, and Sims sent Halbouty a bill for $125. They kept on sending each other bills for several months, but finally stopped because they agreed they couldn't afford the postage. (In later years the two men teamed up on an enterprise that enriched them both.)

Halbouty was reeling under hard blows, but Sims and others noticed that he had not lost any of his basic self-confidence, nor had his bellicosity been diminished a whit. Indeed, his temper played a large part in getting him out of poverty's rut. A New Mexico oil operator, Ray Talmadge, came to the office to employ Halbouty to evaluate some oil acreage he intended to acquire in a neighboring county.

"What will the job cost me?" Talmadge asked.

"Five hundred dollars," Halbouty said.

"Five hundred dollars! For Christ's sake! I can go across the street and get a high-powered geologist to do the job for seventy-five!"

"Get the hell out of here and get the Goddamn thing done across the street!" Halbouty shouted. "I don't give a Goddamn! You'll get a seventy-five dollar report! Get out of here!"

Talmadge left in a major huff. Pearl was almost in tears. "He might have given us enough to pay my salary if you hadn't been so mean. You owe me for two months."

"I'm worth five hundred dollars," Halbouty muttered.

"Oh, sure!" Pearl said.

About twenty minutes later Talmadge returned and Pearl ushered him into Halbouty's cubbyhole. "If you're as smart as you are loud, it might be worth five hundred bucks," Talmadge said. "What do we do now?"

"Give me half of it now and the rest when I give you my report," Halbouty said.

"Fair enough," Talmadge said, and he wrote out a check for $250. Halbouty said he would get to work immediately.

Talmadge was hardly out of the office before Pearl grabbed Halbouty and they danced around the office. "Let's celebrate!" Halbouty said. "Pay me first," said Pearl. "Come on," Halbouty said. He cashed the check, paid Pearl almost all she was due, then they went to the neighborhood sandwich shop and gorged on the house specialty.

After Talmadge received the report and paid Halbouty, he confided to James Clark, oil editor of the Houston *Press*, "That son of a bitch told me more than I want to know about that damned acreage." Clark smiled knowingly. He and Halbouty had been friends from boyhood. When he became an oil editor, he told Halbouty he knew little or nothing about geology. For several nights in a row, Halbouty arrived uninvited at Clark's home promptly at eight o'clock and lectured Clark for two hours. Finally Clark protested. "What the hell are you trying to do, make me the second-best geologist in town?" "No," said Halbouty, "but no friend of mine is going to be the second-best oil editor in town." (Clark later became a top-rank oil historian, and he and Halbouty co-authored several books about the industry.)

All of a sudden Halbouty felt like he had a chance of getting off God's blacklist. He hadn't finished spending the Talmadge fee when he got a call from W. T. Burton, a South Louisiana entrepreneur and political power. Burton had a hand in a dozen ventures, including oil. He and Miles Frank Yount had been friends, and Yount had once "loaned" Halbouty to Burton for a short period.

"I need you, Mike," Burton said via long distance phone from his base in Sulphur, Louisiana.

"I'm a consultant now, Mr. Burton," Halbouty said quickly; he wanted it known he wasn't on loan anymore.

Burton was aware. Johnny Chance, Halbouty's partner in the now-defunct Merit Oil Company, had found a job with Burton. "Johnny told me how to get hold of you," Burton said. "I'll put you on a retainer. If you work, fine. If you don't, I'll still pay you so much a month."

"How much is 'so much,' Mr. Burton?"

Burton laughed. "How's three hundred a month?"

"I'll see you in the morning," Halbouty said.

Sulphur was a small town and Burton kept his office in the rear of a country grocery store. He was signing a big stack of checks when Halbouty entered, and he bade Halbouty be seated. As he signed checks, he snarled at his clerk, a white-haired man named Maxwell. Maxwell was searching the files for a letter Burton wanted. "Find it, damn it!" Burton would say without raising his head.

A kitten leaped into a chair, then made its way to the top of Burton's big desk. It moved close to Burton's hand. Burton shoved it away. The kitten went back. Burton shoved it away, still signing checks, still grumbling at Maxwell to find the letter. "You'd lose your ass in those files if it wasn't tied to you," he grumbled. He shoved the kitten away again. Halbouty was fascinated. "Find the letter, Maxwell!" Burton growled. The kitten extended a paw. Burton grabbed it and flung the kitten toward Maxwell. "File that Goddamn cat away and it'll be lost forever!"

Halbouty laughed. (Through the years the incident would be one of his favorite funny stories.)

Burton had plenty of work for Halbouty in Louisiana, both in established fields and wildcat areas. So much that Halbouty was able to re-employ Milton Cooke, who had returned from Venezuela. Cooke had become a fine petroleum engineer. As Halbouty had instructed him in the past in the days with McCarthy and the Merit Oil Company, Cooke was now able to bring Halbouty up-to-date on techniques he had learned in South America.

About that time Halbouty hired his younger brother, Jim. Jim had not gone to college. Upon graduation from high school, he began selling stocks and playing the violin in Beaumont dance bands. He was such a superb violinist that the Houston Symphony Orchestra lured him to Houston. Halbouty gave him a part-time job of washing well samples for micropaleontological determinations in the office laboratory, an undemanding occupation.

One Friday noon Halbouty was called away from the office. He was working on an oilfield map, trying to correlate logs of wells already drilled in an effort to find a fault he was sure existed. He left his work on his desk. When he returned to his office the following Monday, Jim had correlated the logs and had found the fault. Halbouty was delighted. "Throw away that Goddamn fiddle!" he shouted. "You're going to be a geologist!"

Jim went to the University of Texas, not Halbouty's beloved Texas A&M, obtained his master's degree in geology in four years, and went to work for Standard of Texas with a Phi Beta Kappa key dangling from his watch chain. "He's a Goddamn genius," Halbouty would boast. "You've wrecked his life," Jim's musician friends would counter. Jim, a mild man but not a meek one, would keep his own counsel.

Halbouty, meantime, had achieved notice for finding new production on the east flank of the old Jennings Salt Dome for Burton, and following that with finding new production on the east and northwest flanks of the Starks Salt Dome, another established Louisiana field. And for a group of Kentucky investors he found new production at the Lafitte field.

And he was not forgotten by Ray Talmadge, the New Mexican he had charged $500 for acreage appraisal. Talmadge recommended Halbouty to a New York investor named Martin Walker. Walker wanted, immediately, an appraisal of certain oil leases near Corpus Christi, Texas. He spoke even louder and more rapidly than Halbouty, and the latter failed to name the price for his work in the brief conversation.

But it must be assumed that Walker was pleased, for he paid Halbouty $15,000 for his report. Plus expenses of $1,030. After all the failures and the sudden rebirth, the $15,000 meant more to Halbouty than the $125,000 Humble had paid him for the Friendswood lease when he still had known only success.

After that his fortunes improved almost daily. He began traveling to oil patches all over the country; banks in New York, Chicago, Dallas, San Francisco, and Houston employed him to make evaluations of acreage in the various fields. He had become an expert on the geology of the Gulf Coast. Now he studied as he traveled. He was regarded in his profession as an expert on most of the oil provinces of the country, and he was paid accordingly.

He had begun as a consultant by asking for, and occasionally receiving, as much as $25 a day for his services. As the months passed he increased his fees to $50 a day, then to $150, and eventually to $300. With such fees and several retainers, he was grossing $10,000 a month as 1939 drew to a close.

He still was finding time to write, not only for oil trade publications but for scientific journals as well. Competitors gossiped that he wrote for the oil trade publications to draw industry attention to him. This was probably true. He was aware of the virtues of publicity in getting business. That was one thing he had learned from McCarthy, who apparently believed that even bad publicity was better than none.

But Halbouty's ego also was involved. He loved seeing his name in print, and he was convinced he was a master of the language. He was proud that the "yearbook" of which he had been editor-in-chief had won All-American honors for Texas A&M in 1930. It must be said that his writing had improved steadily. For one thing, he knew that many practical oilmen were not well educated. Therefore, he quite often reduced scientific terminology to simple English and wrote as if he were explaining his conclusions to his barber. Further, he had become aware of the short declarative sentence, and he practiced it until he felt safe in employing more complex structures.

And he attracted attention to himself because his writings sometimes created controversy, something he did not seek at this stage in his life. It came because he had tossed equivocation aside; he wrote his conclusions boldly, for all to see. His 1937 article on the Hastings field—the one that had attracted the overalled Charley Alexander to his office—angered officials of Humble and Stanolind, the companies that controlled most of the field's acreage.

Taxing authorities valued a field by its estimated reserves. Estimating reserves was an inexact science at best at that time, and those who owned reserves felt it prudent to estimate them conservatively. Governments had no scientists of their own in that period. They were required to pursue any rumor or report on reserves since they never quite believed oil company estimates, which also may have been prudent.

Halbouty's article placed Hastings' reserves at more than 520 million barrels of oil, a wildly high estimate by almost any yardstick, including Humble's and Stanolind's. He also wrote that the field's limits eventually would cover more than six thousand acres. And he was so bold as to pinpoint the average acre recovery at 112,543 barrels.

Taxing authorities greeted the article gleefully. They pointed to Halbouty's credentials. They raised taxes on the field even as Humble and Stanolind officials fought to stay their hands. (Hastings field, by 1978, had yielded more than 600 million barrels of oil and was still going strong.)

Halbouty also was in demand as a speaker before geological and petroleum engineering groups. He was a passionate man, and he could fill their disciplines with romance and dedication. To Halbouty, the old wildcatters and creekologists were more colorful—and more American—than the cowboy of the Old West. Standing on a platform, his audience before him, he could instill drama into the boring of a lonely exploratory well and high adventure into the discovery of a well-hidden fault system. This was the chocolate into which he dipped the new ideas or theories he wanted to display.

Alexander Deussen, Mellie Esperson's geologist who had worked with Halbouty at Cedar Bayou, looked on the young man as a future leader of the oil industry. Deussen had been president of the American Association of Petroleum Geologists and had won the group's highest honor, the Sidney Powers Memorial Award. Halbouty, he told his peers then, would someday win the award when it would be even more meaningful.

Halbouty was thirty, and no longer a youngster to those with whom he worked in all phases of the oil industry.

He was able, now, to build a home for Lesly in an exclusive area of Houston, Tall Timbers, and have nurses attend her in his absence. The home's hardwood floors were uncovered, and there were no hindrances to her passage about the house in her wheelchair.

He was rich, respected, and loved, and he still was not completely happy. He sometimes felt as if he were marking time. He had proved over and over again that he could find oil—for others. He would never be fulfilled until he could prove that he could find it for himself.

five

As AN OFFICER in the Army Reserve, Halbouty was called to duty for three months in 1940 at Fort Sam Houston. It was a refresher course for those who had been active in ROTC in college and who had maintained their positions in the Reserve. They were told they would be among the first called to active service should the United States be drawn into the European conflict.

It was a period of intense study for Halbouty, but he remembered it best, and bitterly, too, because his beloved Texas A&M football team was defeated in the last game of the season by the hated Longhorns of the University of Texas. The Aggies had won the football National Championship in 1939 and had defeated Tulane in the Sugar Bowl. They had won all of their games in the 1940 season until meeting the Longhorns and had been promised a trip to the Rose Bowl. But the unheralded Longhorns, a decided underdog, won the game 7-0.

Halbouty, in uniform, had driven up from Fort Sam Houston in San Antonio to see the game in Austin. Not even Notre Dame graduates are so unashamedly sentimental about their school as are ex-Aggies. Halbouty literally was bowed with grief as he left the stadium. L. T. (Slim) Barrow, a Humble geologist and official, slapped him on the back. "Buck up, Mike," said Barrow, a University of Texas graduate. "You may win it next year."

Not seeing who had spoken, Halbouty whirled like a fighting hound. Barrow stepped back, laughing. "Come on, Mike. It's only a game."

Halbouty immediately was ashamed of his quick anger, and apologized. But Barrow and others told the story for years to illustrate Halbouty's temper.

With his "refresher" service completed, Halbouty returned to Houston and his consulting business. He was convinced that the United States would enter the war eventually, and he planned for it. He knew he would be an infantry officer, so he studied infantry tactics in his spare time. And once a week he found an opportunity to spend an hour at a private firing range, where he sharpened his skill with both handguns and rifles.

It was his habit to work on Sunday afternoon, and he was in his office on December 7, 1941, when Lesly called to tell him about the Japanese attack on Pearl Harbor. Forty-eight hours later he was called to active service and ordered to report to Fort Benning, Georgia, on February 12, 1942.

Many of his friends in the oil industry urged Halbouty to "stay out of the army." Some thought he was needed more on the home front. "All you have to do is sign that little paper saying you're in a vital war industry," they told him. Others thought he should stay out of the army because only a fool would go if there was a way out.

He ignored them all. He felt he owed the country a debt because his immigrant parents, poor and uneducated, had found sustenance in it. He felt he would be a good infantry officer, and he believed good infantry officers would be in short supply. In a different vein, his male pride demanded that he soldier, and promised him shame if he didn't. And, a war was an interesting phenomenon, a global war even more so, and he couldn't allow himself not to be a part of it. No doubt many Americans with the choice to go or stay were impelled into service by similar motives. A fellow geologist provided another motive for Halbouty. "The son of a bitch," he quipped, "doesn't think they can run the war without him!" The line produced chuckles at oil-folk gatherings for months.

Halbouty had been appointed a Second Lieutenant in the reserve on his graduation from college. He was promoted to First Lieutenant in 1934, and then to Captain after he went through the refresher course at Fort Sam Houston in 1940. So it was as a Captain that he reported to Fort Benning.

He was in the first class of the war, receiving training with other reserve officers and, to his surprise, a preponderance of West Point graduates. He was not surprised that he graduated at the head of his class. His studying had paid off.

He wanted to ship out to an infantry regiment, but he was made an instructor. He had made up his mind that as long as he was in the army, he was the army's to do what it would with him. So he accepted his lot, or tried to.

Then he received a "Dear John" letter from Lesly.

Because of her physical condition, she wrote, they had not been true husband and wife since early in their marriage. She had been a hindrance to him, the letter said, and was more of a hindrance with each passing year. She loved him, but the love she could give him was not enough for the kind of man he was.

And there was something else. She longed to be with her people; she loved them and was loved by them, and she wanted to spend her days with them in Fort Worth.

She wanted a divorce. She suggested incompatibility as

grounds; it was a socially acceptable claim and would cause neither of them harm. She did not suggest that there had been other women in his life. If she had, he would have denied it. He would have been lying, but he would have denied it. (He would deny it now. But there had been a woman from whom he obtained a lease in Jefferson County; she was still keeping a newspaper scrapbook of his exploits in 1978. There had been others, two of whom reminisced about him fondly as this book was being written.) He had been discreet, and the women had been discreet.

Halbouty read the letter a dozen times enroute to Houston. He had come out of his shock and procured a few days' leave. He was still hurt and angry when he confronted Lesly in their home. He didn't want to be divorced. The government, he told her, would not allow her to divorce him while the war was going on. She listened to his outbursts with the same equanimity she had shown when she refused medical help for her illness. She had made up her mind, she said serenely.

She made no move while he was at home, but after he returned to Fort Benning, she filed for divorce. Halbouty did not contest it. If he were distraught, he did not show it as he went about his duties. But in the passing years he would grow misty-eyed when he spoke of her. (And a look at his financial records in 1978 showed that the first check that left his office each month after the divorce went to Lesly.)

He was now Major Halbouty, and he wanted action more than ever. An opportunity quickly arose. Colonel Robert Foster, a friend, was ordered to form a regiment, and he wanted Halbouty as commander of his first battalion. Everything was being arranged, he told Halbouty. Three days later Halbouty received Presidential orders to report to General William Covell in Washington, D.C., for an interview.

He didn't want to go. He felt that he had demonstrated

leadership qualities, and he wanted to be with the troops. "Don't fill my slot," he told Colonel Foster. "I'll be back." Foster was dubious.

In Washington Halbouty learned that an Army-Navy Petroleum Board was being organized. One post called for an Army officer who had been both a geologist and petroleum engineer as a civilian, and who had never been employed in those capacities by a major oil company. The filing system had spit out Halbouty's name; General Covell, of the Quartermaster Corps, was to determine if the candidate was suitable.

General Covell was jovial. "Well, young man, we're glad to have you here."

"I'm not glad to be here, sir," Halbouty said.

Covell's joviality evaporated. "Tell me why," he demanded.

"I think there has been a mistake, sir. I've been trained as an infantry officer, and I think I'm a good one. I've got an opportunity to be a battalion commander, and I want to go back to the troops."

The general's voice was icy. "You haven't given me an opportunity to tell you what we want to do with you."

"It doesn't make any difference, sir. I still want to go with the troops."

"It doesn't make any difference what you want," Covell said. "We'll tell you what you're going to do."

"What *am* I going to do, sir?"

"I'll let you know when I'm ready, Major. You go back to Fort Benning and straighten up your affairs. Get back here. And I don't ever want to hear another word of argument from you."

What he did was sit around and wait. He was destined never to hear a shot fired in anger, and the only troops he saw were the enlisted men who saluted him on the street or as he crossed some hotel lobby. He rented an apartment, and eventually the Army-Navy Petroleum Board was put together. His job was in the Petroleum Production Section. Because he seemed to do everything well, he was promoted to Lieutenant Colonel and became chief of his section. He reported to an admiral, who reported to the Joint Chiefs of Staff.

His job was to increase oil production to sustain the war effort. The Joint Chiefs had wanted an independent oilman for the job—someone free from major oil company influence—because it was necessary to study the major companies' records to learn their production practices. The majors wanted it that way also. They didn't want a competitor's man prowling their books.

Oil from United States fields was powering the Allied war machine. The giant East Texas field alone was supplying more than a hundred million barrels a year, and other fields were producing to their limits. Steel allocations were coveted, and oil companies had to vie with other industries and each other for the pipe and equipment needed to drill more wells to produce more oil.

But pipe and equipment also was needed in the oil-producing countries of South America that were friendly to the Allied cause. Halbouty was sent to the Caribbean oil province— Venezuela, Colombia, and Trinidad—to see what could be done to raise production. As a result of his studies, supplies were diverted from United States fields to the Caribbean and production in that area was more than doubled.

In Bogotá, Halbouty became a public figure by virtue of innocence and luck. In uniform, he left his hotel room one morning, briefcase in hand, to visit the U.S. embassy, where he had an appointment with an attaché who was the U.S. expert on oil and minerals. When he reached the hotel lobby it was crowded with noisy Colombians. Halbouty's mind was on his business and he pushed his way through the crowd.

They grew silent at his appearance, and they stared at him as he absentmindedly plowed through them.

The sidewalk and street outside the hotel also were packed with a noisy throng. Halbouty paid the Colombians no attention. He set out for the embassy, a couple of blocks away. He was not aware that the Colombians were peering after him in amazement.

He was about half a block from the embassy when he noticed that armed soldiers were standing guard at the gate. At that moment he saw the attaché running down the walkway from the embassy to the gate. "Run!" the attaché shouted as he raced along the walkway. "Run!"

Halbouty did not run. The attaché shouted, "Goddamnit, they're demonstrating against the United States! Hurry!"

Halbouty walked up to the gate. The attaché was furious. "You could have been killed! They've been harassing Americans all night!" Suddenly he burst out laughing. "They couldn't understand why you weren't afraid, Mike! They were too surprised to do anything to you!"

"I'm afraid now," Halbouty said.

He spent the day and the night in the embassy. The next evening, order was restored. And the following afternoon, wearing borrowed clothing and using borrowed clubs, Halbouty shot a hole-in-one on the seventeenth hole of the Bogotá Country Club golf course, using a five-iron on the 165-yard, par-three hole.

"*Uno!*" his caddy cried, and other caddies around the course took up the cry.

"Let's get out of here," said the attaché.

"Why?" Halbouty asked.

"You'll have to stay here the rest of the night and buy drinks for everybody! It could cost you four or five hundred dollars!"

"I haven't got but about forty bucks," Halbouty said, "but I'm damned sure going to buy drinks with it. Look . . . I walked through a demonstration and I've shot a hole-in-one. I'm not going to sneak out of anywhere!"

So he put his uniform back on in the locker room and spent his forty bucks buying drinks in the clubhouse. He showed his empty pockets. "No more *dinero,*" he said.

The Colombians were impressed by his lack of pretense. Whooping, they set upon him and bought *him* drinks. He left the club a hero, and was so greeted on the street and at his hotel, for word had spread quickly that the North American with the stout heart also had performed an incredible feat with a slender club and a small ball.

Back in Washington, he met a striking divorcée, a strawberry blonde with gray eyes and an Oklahoma accent. She was Fay Renfro Kelly. She worked for the American Red Cross and had a ten-year-old son, Tommy, who was attending nearby Staunton Military Academy. Halbouty courted her with an ardor he had never known before. She was a gourmet, and loved to dance. Halbouty needed instruction in wining and dining, but he had learned to dance on the rough floors of oilfield honky-tonks, and he slid onto the polished floor of the Shoreham Hotel's Blue Room like an otter returning to the sea.

His courtship was cut short by his superiors. He had done a good job in the Caribbean oil province. Now they sent him to the remaining oil-producing countries of South America. Again he impressed them with a fine performance.

In Santiago, Chile, as he browsed through a small shop, his eye was caught by a strangely shaped bottle of Chilean wine. Fay, he thought, would love to have a wine so rare and unobtainable in the United States. He bought three bottles, paying $2.50 each. He carried the bottles with him for more than twenty thousand air miles, wrapping them tenderly before each flight in a thickness of three or four towels.

The bottles were intact when he returned to Washington. He hosted a party in his apartment, and at the proper moment unveiled the wine. The unveiling was accompanied by an exaggerated account of the perils he had encountered while trying to preserve the precious beverage for his friends.

The next day they went to a liquor store to buy a bottle of Scotch, and Fay led Halbouty to the wine racks. The first wine he saw was in a peculiarly shaped but familiar-looking bottle—with a Chilean label—priced at $1.75!

"Don't you say a Goddamned word," Halbouty warned her. "I've learned my lesson."

She smiled. "It's my way of saying that you don't have to be anything but Mike Halbouty to me."

His superiors sent him to an old stomping ground, South Louisiana. The government owned a canal and the right-of-way through which it was cut in the Lafitte oilfield, an old field where Halbouty had discovered new production for some investors. Both independent and major companies were trying to buy drilling rights on the property. Halbouty's job was to appraise the acreage for its oil value and thus give the government an idea of its lease worth.

He set up shop in New Orleans. With his departure from Washington, Fay had gone to Los Angeles to visit a sister. Halbouty spent a great amount of his army pay on long-distance telephone courting.

He interrupted his work at Lafitte field to conduct some studies for the government near Wichita, Kansas, and Tulsa, Oklahoma. Tulsa was Fay's hometown. He called her from there one night. "Goddamnit, I miss you. Let's get married."

She met him in Newkirk, a small town near Tulsa, and a Justice of the Peace rustled up some of his friends to witness the wedding ceremony. He took her with him to New Orleans and, after a proper interval, Tommy joined them for the summer. The three of them were walking down Canal Street when newsboys shouting "Extra!" brought them word of the atomic bombing of Japan.

Halbouty handed Fay the newspaper after he had read the story. "Well," he said, not entirely facetiously, "your new husband is going to have to learn a new profession. We've just left the Liquid Fuel Age and entered the Atomic Age. That's where energy will come from from now on."

She looked at him questioningly.

"Not right this minute," Halbouty said, laughing.

He was bored, waiting for the war to end, waiting to be a civilian again. He went to see some old friends, Gage (Bud) Lund and Ken Crandall. Lund was general manager of the California Company, a subsidiary of Standard Oil Company of California, and Crandall was his chief assistant. Both men were geo-scientists.

"Give me some logs to fool around with," Halbouty said. "Anything. I've got to get my hand back in the game."

Lund laughed. "Well, we just drilled some dry holes in Natchitoches Parish. You can have those logs. You can have the maps and everything else. We dug seven holes to eight thousand feet and didn't find a whisper." He laughed again. "Nobody's ever found anything in that parish. We're through looking, and so is Shell."

So Halbouty took the material home with him to study, just to get his hand back in the game, as he had said. He needed the practice. But after several days of study, the practice became exciting work, because his eye had found something of worth. Both the California Company and Shell had sent their geologists over a vast area in the parish, and both companies had sent seismograph crews in to "shoot." Neither the geologists nor the geophysicists had found a deep structure they obviously were seeking. Halbouty's eye found a shallow structure, or so he thought.

He went back to Lund and Crandall. "Look," he told them, "I think you've got a prospect here."

Lund shook his head. "Nope. We couldn't find anything, and we did a lot of shooting."

"I'm looking at it strictly from a sub-surface point of view," Halbouty said. "I think there is some shallow stuff there, maybe in the Tuscaloosa sand."

"We're through with it, Mike," Lund said.

"I think it ought to be drilled for shallow production," Halbouty insisted.

"Not by us," Lund said.

"Well, can I work out something with you? Will you farm it out to me?"

"Hell, yes! But what are you going to do about the other acreage? You need a lot more."

"I'll work that out with Shell."

They shook hands on the deal.

Still in uniform, Halbouty went to the Shell office in New Orleans. Shell, too, had given up on Natchitoches Parish; the company was happy to farm out its acreage to Halbouty. With both companies the formula was the same: the companies retained a sixteenth interest, the land owner two-sixteenths, and Halbouty thirteen-sixteenths.

It was a handshake deal with Shell, also. "When I get out of service, I'll be back to sign the papers," Halbouty said.

He was detached from army service in September 1945. He made a quick trip to Calgary because he thought Canada was floating on an ocean of oil and because he had promised himself a look at it. He arrived in a snowstorm, spent two days in a snowstorm, and left in a snowstorm. Canada, he decided, was not for him. Houston would be his home base.

He bought Fay a house in the exclusive River Oaks section with a small down payment. He went to Natchitoches Parish and acquired leases on land he felt he needed to complete his drilling block. He formalized his deal with Shell and the California Company. And when he returned home he was broke.

He walked the streets of Houston trying to sell three-quarters of his working interest to get drilling money. No one believed in the Natchitoches Parish area because of the dry holes drilled there. He explained that he wanted to drill north of the dry holes, that he was certain of finding shallow production. Oil company officials told him he was making a mistake. Bankers gravely shook their heads no. But, he insisted, a shallow well would cost no more than twenty thousand dollars to drill. Was that too much to gamble on a geologist with his reputation? The Shell geologists had good reputations, they pointed out to him, and so had the scientists with the California Company.

Finally a fellow geologist, Ken Owens, agreed to put in ten thousand dollars, all he could afford. "I'm tired of seeing you walk the streets," he kidded Halbouty. In fact, he agreed with Halbouty's geology.

But Halbouty still needed ten thousand dollars. What he did to get it was the beginning of his career as an oil operator of the first rank. He sucked up his nerve and went to New Orleans to see Lund and Crandall of the California Company. He showed them the geology he had worked out; they had not seen it before.

"I've put this block together," Halbouty said, "but I can't promote the drilling money. I need some help with the son of a bitch."

"What the hell do you want us to do?" Lund asked. "We practically gave you the acreage."

"I want to sell you an interest in the prospect," Halbouty said evenly.

Lund and Crandall looked at him as if he had lost his mind. "I never heard such unmitigated gall in my life," Lund said finally. "Asking us to put real money in something we practically gave away. You ought to be ashamed of yourself."

"Damn it, your company will get a bigger interest if you buy in now than that little override you've kept. When I hit you guys will be heroes to your bosses, and I'm going to open a new field—and don't you forget it!"

Lund looked at Crandall. "You know, Ken, any son of a bitch who's got the nerve to ask a company to join in drilling acreage it farmed out in the first place ought to be helped. By God, I'm going to do it!"

He had gone for himself in 1937 when he left Glenn McCarthy, and he had failed. His ability to find oil for others as a consulting geologist and petroleum engineer had done little to soothe his badly bruised ego. Now, in December of 1945, he was getting a second shot at the moon.

He had all the incentives, including a financial one. He was broke. He was in debt. But on the brighter side, the price of oil was going up. The federal government had frozen the price of crude at $1.25 a barrel during the war. But the postwar demand for automobiles and the fuel to run them already had driven the price of crude to $2.00 a barrel, and anyone with an eye to see could calculate that $3.00 and more per barrel was at the next bus stop. All Halbouty had to do was find a field.

He dealt with a drilling contractor whose crew spudded in the well on the C. L. Detro farm. Thus the well became the Detro 1. The acreage was near the village of Ashland, but Halbouty and the drilling crews set up housekeeping in a motel in the small town of Coushatta. Fay was with Halbouty, and on weekends she went to Houston to bring Tommy back with her from the private school he was attending there.

Halbouty paced the rig floor at the Detro 1. The roughnecks called him "Colonel" because he wore his military uniforms as work clothing. They called the well a "cheapie" because Halbouty had assured them they would find the Tuscaloosa sand at above 3,500 feet. They thought Fay was "the genuine article" because she spent almost as much time at the rig as did Halbouty.

There were few visitors to the drill site. The oil fraternity considered it a condemned area, and the residents no longer were curious about clanging equipment that did little to enrich them. The weather would have discouraged most visitors anyway. It was cold and wet and the stinging wind whistled through the derrick like a lost freight train.

It would have taken an earthquake or a hurricane to dim Halbouty's enthusiasm. His exhortations amused the crew. "That cheer-leadin' son of a gun could have *talked* the Germans and Japs to death," one roughneck told another in Fay's hearing. At another time she heard some crewmen discussing Halbouty's penchant for sampling cuttings from the bore hole. "I'll bet, by God, he's et a *ton* of dirt in his lifetime," a roughneck vowed.

The drilling went smoothly, and even the weather straightened up for Christmas. Christmas was a work day, for the drilling could not be halted, but it did not produce the present Halbouty was expecting. Fay, however, made the day memorable. Two roughnecks built her a serving table, and she played hostess at a buffet supper that was brought in by truck from Coushatta. Also, a little whiskey drinking was done.

Halbouty had ordered that cores be taken intermittently when the drill reached 2,900 feet, the shallowest depth at which he thought the Tuscaloosa could be encountered. He presided over the examination of each core, and each one was a disappointment. "If this bullshit keeps up," a roughneck murmured, "he's gonna lose his zip."

Three days after Christmas a core was taken below 3,280 feet. When the core barrel was removed from the hole, the driller, a tall man, noticed bubbles rising in the mud clinging to it. "Gas, Colonel, sure as I'm a foot high!"

The core barrel was opened. It had cut 16 feet of material,

and 8½ feet of it was oily Tuscaloosa. "Great God from God-ville!" the tall driller said.

Halbouty was strangely silent. He rubbed the oily sand between his fingers. Finally he said, "We haven't produced it yet, boys."

"Oh, hell," said the tall driller. "It'll make a well, Colonel."

Halbouty grinned at him. "Sure as you're a foot high!"

The well was ready to bring in the night of December 31, and Fay had a bottle of champagne to break on the rotary table, but Halbouty held back. Since the well would be the first ever in Natchitoches Parish, he had arranged to notify the Associated Press office in New Orleans when the flow began. But he wanted to milk the maximum publicity from the event. "We'll bring her in one minute after midnight," he told Fay and the crew. "It's going to be a New Year's baby, the first well in the State of Louisiana in nineteen forty-six. The first well and the first field."

And that's the way it was. The Detro 1 came in at one minute past midnight, flowing 120 barrels of oil per day through a 9/64-inch choke, to be officially listed as the discovery well of the Ashland field.

With the well in, the Halboutys and the crew drove into Coushatta to celebrate the well and the New Year. The sounds of merriment shook the motel. Soon a few villagers dropped by to see what was going on. Then others arrived. Eventually every adult with a thirst or curiosity or both showed up. Toasts were drunk to the "Colonel" and the new prosperity. Standing on a toolbox, Halbouty made a speech accepting the praise being heaped on him.

If he seemed to regard his listeners as combat infantrymen, it could be blamed on the booze, his love for foot-soldiering and the way he was being "Coloneled." "Men," he cried, encompassing the women present in his wide gestures, "we've just whipped hell out of that son of a bitch out there, and by God we're going to whip some more!" He went on some more in that vein, but daybreak found them all back at work at the well.

And four months later, to the day, in the adjoining parish to the north, he discovered the Lake End field with his Savill 1, a producer that was drilled to only 1,260 feet.

He sold out his interest in both fields, and was ready to move on.

"He's glad to have the money," Fay told a new Louisiana friend, "but he's happier, I think, that his geology proved out. His pride needed these discoveries more than his pocketbook."

six

FINDING TWO OILFIELDS within four months was like hitting a home run on a 3-2 pitch in the seventh game of the World Series, bottom of the ninth inning, two outs, the bases loaded, and his team three runs behind. The odds were 9 to 1 against striking oil with a wildcat well in the United States, 16 to 1 against finding a very small field, 53 to 1 against finding a small field, 330 to 1 against finding a medium-sized field, and 991 to 1 against a major strike. The one in ten successful wildcat well, then, ofttimes located an oil pool that could not produce in truly commercial quantities.

Halbouty's feat was scarcely mentioned in the newspaper oil pages and the oil trade press. He was disappointed. Further, during his time of failure before the war, and during the war, he had seen pages of stories about Glenn McCarthy's strikes and growing wealth. He was always aware of McCarthy, and he believed McCarthy was aware of him. Halbouty felt that he was being measured against the other man.

Still, he did not sashay out immediately to find a new field. Instead he made a deal with Del Cryer, a Shreveport drilling contractor who had drilled the first wells in the Lake End field for him. Cryer knew that some leases could be obtained near the old Pine Island field in Caddo Parish. If Halbouty would buy the leases, Cryer would drill the wells at cost. From there they could arrive at an equitable distribution of profits, if profits there were.

Halbouty had read the literature on the old field, and he drove to Caddo Parish to have his look at it. It was a different kind of field. While most fields produced from loose sands where the oil could migrate and accumulate against a fault of some kind, the oil in Pine Island field came from the Austin Chalk, a soft, gray stone composed of shells of ancient marine creatures. The oil, then, was embedded in what amounted to solid rock, and the minute globules were difficult to seduce to the surface. But, as Halbouty and others before him had decided, there was a hell of a lot of oil trapped in the chalk.

Halbouty reckoned that the acreage Cryer had access to held a hundred thousand barrels to the acre but could produce only four thousand barrels to the acre. Nonetheless, he leased the land available. The risk was not great. The oil normally was found at fifteen hundred feet, which meant a well cost twelve thousand dollars to complete.

The first wells came in flowing seventy-five and eighty barrels a day, dwindled to twenty-five barrels, then settled down to five or six a day. Halbouty deduced that they could produce that much far past his lifetime. The answer, then, was to drill many wells. A hundred wells could produce five hundred barrels daily in perpetuity. At roughly three dollars a barrel, that amounted to fifteen hundred dollars daily. "Annuity income," it was called in the industry.

But on the ninth well, just past one thousand feet, the bit and the drill pipe suddenly dropped twelve feet, as if into a cavern. There was a show of oil, and indications that the well was ready to blow out. The crew pumped in mud to hold it back. Casing pipe was set. The well was brought in, and it came in with a roar.

They couldn't shut it off. The master valve on the wellhead assembly wouldn't function. Oil was flowing from the well at a rate of more than twelve hundred barrels per day, and Halbouty had only two small tanks to hold it. They were filled in less than an hour. The driller and a roughneck struggled with the master valve. Halbouty sent a roughneck in a pickup truck to find another valve. Then he, Cryer, the rest of the crew, and some farmers pressed into service went to work with picks and shovels. They cut trenches to run the oil into natural depressions. A crew from a pipeline-gathering system began laying a line to the well.

"I will always remember Fay the way she was at the well," Halbouty wrote his brother, Jim. "The terrain is swampy, as you know, and every damned mosquito in Louisiana was trying to eat us alive. But there was Fay. She had wrapped newspapers inside her clothing and her stockings to ward off the mosquitoes, and she was standing ankle-deep in ooze, swinging a shovel as good as any roughneck . . . "

The well flowed for three days before a new master valve was set in place and the well shut in. About three thousand barrels of oil were pumped from the depressions and ditches. The well continued to produce at a high rate for months, decreasing its output slowly until it reached the five-barrel-a-day average. It had paid for itself with the first week's production.

The twelve-foot cavern in the well gave Halbouty an idea; he would create his own cavern. Some operators at Pine Island made a practice of pouring several barrels of hydrochloric acid into their wells. The acid ate the chalk and forced the release of more oil. Halbouty had his drillers log every "soft spot" as they spun the bit downward in the next well. Casing was set and perforated at each soft spot. Then a well-service company pumped hydrochloric acid into the well under great pressure. The acid went through the perforations in the casing and ate into the chalk, creating five great caverns at different levels.

The well came in flowing two thousand barrels of oil per day, and this time the tankage was adequate. Word of what Halbouty had done raced through Pine Island field. For the first time in decades, there was a rush for leases. But Halbouty, gambling that this theory would work, already had obtained all available leases in his area.

His acidizing method became a common practice in the field. Over a period of years he would drill ninety-eight wells on his leases, buy out Del Cryer, and eventually sell out himself because he didn't want to tie himself to what he considered a marginal venture. (In the late 1970s he would regret the sale; the wells were categorized as "strippers" because each produced less than ten barrels a day, and "stripper" oil was selling for about fourteen dollars a barrel! The ninety-eight wells brought their owners about six thousand dollars daily.)

He spent little time in the Pine Island field area once he had production started. With this "annuity income" established, he struck out looking for a big one. His steps took him to the old South Liberty field on the Trinity River, about midway between Houston and Beaumont.

Meanwhile, he also established a postgraduate fellowship in geology, in his name, at Texas A&M, awarding five hundred dollars annually to a student who had worked while in school obtaining his bachelor's degree. And he hired a young geologist.

When Fay heard this news, she cried, "You're out of your mind, Mike! You're practically broke. You've got everything you've got invested in Pine Island. Sure, you have the five hundred dollars this time . . . but what about next year and the other years? And how are you going to pay this geologist? You can hardly pay your secretary now."

Halbouty said something to her then she would many times have occasion to remember. "I'll always be just one step ahead of whatever income I have, Fay. I think I'm going to make millions, but I'll lose millions too, because I'll always be going out where others won't."

Why did he establish the scholarship at such a time? He loved the school, yes. He remembered his own difficulties, certainly; the work he had done to get his education had been monumental. And there was a broad but often undetected streak of impulsive generosity in him. And there was his ego, his craving for recognition, his love of seeing his name in print. It was a mix—a splendid one, perhaps—that would prompt him to more than a score of public acts of generosity, many times when he could not afford them. ("If Prexy Walton hadn't loaned me fifty dollars, and if I hadn't worked the shirt off my back to pay him back, I'd still be selling bananas in Beaumont" would be a refrain throughout his life.)

The geologist Halbouty hired was George Hardin, Junior, twenty-six to Halbouty's thirty-six, not long out of Texas A&M, not long on experience, but with a stability and brilliance that Halbouty recognized.

At the interview Halbouty saw a stocky, clear-eyed country boy with thinning hair and a calm, confident manner. He had done his graduate work at the University of Wisconsin and had been employed by the U.S. Geological Survey. Halbouty didn't mind Hardin's youth; hadn't he been a boy wonder himself at twenty-six? And he was impressed by Hardin's diamond stickpin, worn with a flourish. But he was worried about one thing. "George," he said, "you don't have any experience at all on the Gulf Coast. What makes you think you can understand Gulf Coast geology?"

"Mr. Halbouty, as far as I'm concerned, I'm a geologist. I

can understand geology wherever I go. All I have to do is stay in that area long enough and I'll understand it."

"You're hired," Halbouty said. "Your salary is four hundred a month."

It was a fine decision. Hardin was Halbouty's equal in many regards, and Halbouty was the first to admit that Hardin's calm judgment was a restraining influence on his impetuous nature. Together they would make themselves and others wealthy and, as Halbouty had told Fay, they would go where others wouldn't go. (Hardin retired a millionaire at forty on the percentage and increased salary he earned with Halbouty.)

Hardin understood what Halbouty was about at South Liberty field. It was an old field, first drilled in 1902, a year after the great discovery at Spindletop. Wildcatters were drawn there because of a vast salt dome. The crest of the dome produced little oil. In 1925 good production was achieved on the east flank of the dome. Through 1946 the field produced about sixteen million barrels of oil, but production had dropped to less than five hundred barrels per day when Halbouty decided to try his hand.

He had been at South Liberty before. Yount-Lee had drilled several dry holes near the crest of the dome in the late twenties, and Miles Frank Yount took Halbouty to examine the area in 1933. Both believed that great quantities of oil could be obtained on the northeast and northwest flanks of the dome, but Yount died before he could move his company in.

Halbouty and Hardin re-examined the area, and Halbouty's initial conviction was reinforced by Hardin's enthusiasm. The owner of the acreage they wanted to lease was Leon Mitchell, a Negro. Mitchell enjoyed Halbouty's sales pitch, but he had bad news. Charles Gainer, a lease broker, had given Mitchell some money and promised to have a well drilled on the acreage. Halbouty went to see Gainer. Gainer was as short of funds as Halbouty, but he drove a hard bargain. He would allow Halbouty to drill the acreage, but he and Mitchell were to share three-eights of any oil produced. Since Halbouty wanted to retain one-eighth for himself, this meant he had to go to a money man with only half of potential production as an offering . . . and money men were accustomed to receiving seven-eighths.

"Unheard of, Mike," he was told time and time again as he went about trying to make a deal.

"Unheard of," said Lenoir Josey, a successful oilman and unsuccessful crapshooter. "You want me to give up three-eighths, carry you for an eighth, and pay for everything. You're crazy as hell!"

Halbouty had reason to believe that Josey might be a bit eccentric himself. He had seen Josey lose $520,000 in a single hour of dice-rolling at the Beverly Club near New Orleans. And now, with Halbouty in his office to discuss the South Liberty venture, Josey was standing by a window and cutting his own hair with the aid of a hand mirror. Chad Nelms, Josey's associate, also was in the office.

Josey acknowledged Halbouty's presence, but motioned with his scissors for Nelms to join him at the window. "Chad, the way this town's growing, that block across the street ought to quadruple in value in five or six years. We ought to buy it. We can probably get it for a million bucks."

Nelms snorted. "Will you listen to this son of a bitch?" he implored Halbouty. "Talking about spending a million and he's so cheap he's cutting his own hair!"

Josey was a warm, friendly, moon-faced man. He associated frequently with such underworld gambling figures as "Dandy Phil" Kastel of New Orleans; Al Smiley, who was Bugsy Siegel's running mate on the West Coast; and others from New York, Boston, and Chicago. Reports had it that Josey was putting underworld money into the oil business, but his friends insisted that Josey simply was fascinated by the hoodlums. He was well liked in the city and belonged to some of the best clubs, including the River Oaks Country Club, where Halbouty had become a member.

When Josey was satisfied with his haircut, they sat down to business. Josey already was incensed that the deal called for him to receive less than he normally could anticipate. Now Halbouty told him the first well likely would be dry. "I want to drill far out on the northeast flank. I'm sure we'll find some sands, but I think they'll be dry out there."

Josey rolled his eyes drolly at Nelms.

Halbouty explained that the early-day production was from Frio sands, but that some thin Yegua sands had been penetrated by the drill. He believed the thin Yegua sands had been separated from main bodies of Yegua sands farther out on the flanks. He pointed out that a Superior Oil Company well on the east flank had found seven Yegua sands but little oil. The thick, productive Yegua sands, his studies had told him, were on the northeast and northwest flanks. "We'll find them with the first well, and I'll know where to get the oil with the second. We're going to hit! It's a cast-iron cinch!"

Josey finally succumbed, but his terms were far from generous. "I'll go along, Mike, but I'm going to pay for your share of the drilling costs on the first well only. If it's dry, like you say it's going to be, you're going to have to pay your share of the costs on the second. And you're not going to get an eighth!"

As the "finder," the geologist and the "ramrod," Halbouty normally would have received an eighth without having to pay any drilling costs. He knew he would have to borrow money to pay for them. And when Josey got through calculating, he reckoned that Halbouty would receive one-twelfth of Josey's five-eighths!

Halbouty accepted the deal; he had to prove that his geology was right.

The first well, on the northeast flank, was as dry as unbuttered popcorn, but it established the presence of four thick Yegua sands and a much thicker Cook Mountain sand. It was a joyous Halbouty who reported to Josey and Nelms. "Do we have the sands!" he shouted. "Do we have the sands!"

Josey shook his head. "By God, Mike, I said you were crazy, and you are. I've seen lots of geologists, but I never saw one so Goddamned excited about a dry hole!" He turned to Nelms. "How you figure a man like that, Chad?"

"I figure we got to drill another well," Nelms said.

As he had anticipated, Halbouty had to scuffle for money to pay his one-twelfth of the drilling cost of the second well. He located the well fifteen hundred feet nearer the crest of the dome. It came in as a fine producer. They quickly drilled four additional wells, with each of the five wells producing from the five separate sands found in the dry hole. They obtained other leases, and on the northwest flank Halbouty found oil in another Yegua sand and later in a Cockfield sand.

By 1950 they had drilled twenty-nine wells with only two dry holes. Halbouty's interest was bringing him about five thousand dollars a month. And Josey had brought into the partnership an interesting mix of solid citizens and some of the state's most notorious gambling figures.

One was Jakie Freedman, who ran an elaborate gambling layout in an elegant mansion near the end of South Main Street. From the mid-forties until the mid-fifties, Houston ran wide-open to gamblers who made payoffs to city and county officials through "Fat Jack" Halfen, a hefty hustler known as the "Big Fix." Freedman boasted that his "fix" ran all the way to the Capitol in Austin. His customers, mostly free-spending oilmen, made him the envy of casino operators from Las Vegas to Monte Carlo.

Freedman joined the venture because Josey had a bad night at the crap tables and didn't have enough cash with him to cover his losses. He suggested that Freedman take an interest in the South Liberty drilling in lieu of payment. Freedman liked the idea and brought in an associate, B. C. (Brother) McKnight. McKnight was the gambling czar of neighboring Fort Bend County, controlling everything from slot machines to a plush casino called "The House of Green Light."

Herman Williams, who had a piece of almost every gambling action in the city, also entered the deal, and so did a policy racket operator named Allison.

(When Freedman's "fix" finally came unglued, he moved to Las Vegas and helped build the Sands Hotel. He is best remembered in that city for losing five hundred thousand dollars in five rolls of the dice at a competitor's crap table!)

When the South Liberty field began shaping up, Halbouty moved out in several directions. One move took him on a sentimental journey. Another would lead him to an unsuspected treasure trove.

He took a train to San Antonio to see Marrs McLean. It will be recalled that it was McLean who induced Miles Frank Yount to drill his leases on the Spindletop flanks in 1925. It was McLean who had sold to Yount the High Island acreage where Halbouty unofficially had discovered the field as a chain-puller. And it was McLean who had deduced that the High Island salt plug was mushroom shaped several years before Halbouty independently arrived at the same conclusion.

Yount had introduced the two men and they had remained friendly over the years. Halbouty had studied McLean's oil career; he had been mystified and amazed at how McLean, with no formal education in geology, had been able to fathom the intricacies of salt dome formations.

In 1920 McLean had bought a five-acre plot of land at Spindletop. It was shaped like a triangle. He was going to drill the acreage in 1925, but Atlantic Oil and Refining Company made him a lease offer he couldn't refuse. He would receive a fifty-thousand-dollar bonus whether or not the lease produced. If it did produce, his bonus would be a hundred thousand dollars plus a twenty percent royalty and fifty percent of the profits.

Nineteen wells, all producers, were drilled on the lease. They yielded millions of barrels of oil, finally went dry, and the lease reverted to McLean. Now here was Halbouty in McLean's office wanting to make a deal.

McLean was incredulous. "Why, Mike? Those nineteen wells sucked that five acres dry!"

Halbouty couldn't tell McLean what was in his heart. He wanted to find oil where his heroes had found it. He had grown up in the shadow of Spindletop's derricks and to find oil there would be a sort of triumphant homecoming. And he could think of nothing more rewarding than to see his name linked with McLean's on a sign denoting a well—Michel T. Halbouty's McLean 1.

More prosaically he explained that he hoped to find some ten-foot stringer sands that might hold enough oil to show a profit. He did not say that both he and George Hardin had some theories they wanted to test. Perhaps McLean sensed that Halbouty was being drawn to the old field by more than a hope for gain. He dealt on Halbouty's terms.

The nineteen wells had been drilled on the triangle's perimeter. Halbouty set up his derrick in the center of the triangle. Cores from the well told an exciting story of the past. From one thousand to four thousand feet the bit passed through alternating sands and shales, with the sands totaling almost two thousand feet. Salt water had encroached on the sands, but each showed traces of the oil that had flowed to the surface by the millions of barrels through the nineteen wells.

Halbouty had been truthful when he told McLean he hoped to find some thin sands the others had missed. He found none. But he and Hardin had calculated on drilling deeper than their predecessors. They did—and the bit plunged into 210 feet of oil-saturated Frio! It was virgin sand; it had never been produced! Had the oil been allowed to surge upward through the open casing as had the Spindletop gushers of old, it would have spouted high above the derrick!

Halbouty quickly acquired other leases, and all proved productive. One lease, a farm-out from Stanolind Oil & Gas, brought him particular satisfaction. The acreage was on property that had originally belonged to the McFaddin family of Beaumont. The first great gusher was drilled on McFaddin land, and so was Miles Yount's discovery well that started the 1925 boom. Now here was Halbouty, a quarter of a century later, reviving the old field. As a boy, Halbouty had delivered groceries to the McFaddin home, and he had never forgotten that he had had to make the deliveries at the back door.

The Spindletop leases, then, paid off in several ways, the most tangible way a financial one. The last of the wells was not depleted until November 1977, after it alone had yielded more than two hundred fifty thousand barrels of oil. It was on the McFaddin acreage.

Halbouty had found oil in areas where salt domes were not present, and would do so again. But he had an affinity for salt domes. He understood them. It is likely no man ever understood them better. He and Hardin were constantly studying old domes, now practically deserted, where oil once had been produced in abundance. He was convinced that oil had been missed in quantity on most of them.

And drilling on salt domes had a financial advantage as well. The Texas Railroad Commission, the oil regulatory body, had devised a "discovery allowable" to fit into its proration plan. A discovery well could produce 20 barrels per day for each thousand feet of depth, plus 20 additional barrels.

Thus a seven-thousand-foot well, like most of those at South Liberty, could produce 160 barrels per day. The discovery allowable applied to the first five wells drilled in the discovery sand. They could produce their 160 barrels daily each day of the month for eighteen months, or until a sixth well was drilled into the formation—whichever came first. Then they were placed on, say, a fifteen-day production schedule at 65 barrels daily.

But each new producing sand was considered a discovery by the Railroad Commission. At South Liberty, Halbouty had found seven separate producing sands. Further, a huge fault split the sands, creating, in effect, fourteen separate discovery zones.

Oilfields off salt domes—those found on anticlines and other structures—seldom have a great number of separate producing sands. The East Texas field, for example, produces from a single stratum of Woodbine sand.

Halbouty set out for an old salt dome near Boling, some fifty miles southwest of Houston in Wharton County. Some wells of no consequence had been drilled on the dome's crest in 1902, but the Texas Company had found good production on a flank in 1923. When Halbouty arrived and obtained acreage on the south flank, it was the only area that had not yielded oil, though wells had been drilled on it.

Indeed, the south flank had been condemned by a number of geologists. They argued that while the Frio—the productive sand—was found at three to four thousand feet on the other flanks, it had been located at from six to seven thousand feet in the dry holes on the south flank. If there had been oil in the south flank, they argued, it had migrated upward into the shallower Frio on the other flanks.

Halbouty rejected the notion. He reasoned that the salt plug had caused such faulting that major migration from the south flank would have been impossible. And he knew that the "high side" on some domes had been prolific, while the "low side" had been prolific on others; he wouldn't condemn a salt dome area because of its structural position.

So he leased some acreage, brought in Lenoir Josey and members of his family—and proceeded to drill three successive wells without finding even a smidgen of Frio. "Anybody with shit in their blood can get out now," Halbouty told his frowning partners.

A fourth well, drilled to 7,213 feet, found 500 feet of Frio—full of salt water. Hardin called Halbouty at home at three o'clock in the morning with the news. They cursed and commiserated with each other. But they decided that the oil they sought was nearer to the crest of the dome and at a shallower depth; the Frio there would hold oil, not salt water. How much was brilliant deduction? How much tough confidence? How much their optimism?

"We ought to pull back up to about three thousand feet and sidetrack that rascal," Hardin said.

"Exactly!" Halbouty said.

Sidetracking meant drilling a directional well with the aid of a whipstock, a device that forced the bit from its vertical path, sometimes by as much as sixty degrees.

"How about the partners?" Hardin asked.

"I'll handle them."

"How about the money?"

"A certain major oil company will guarantee it," Halbouty said.

Humble, aroused by Halbouty's interest in Boling, had acquired an adjoining lease to the west. Halbouty, selling like he had never sold before, went to his partners and was assured of their continuing participation . . . if. The "if" was Humble. So Halbouty went to Humble as he had planned to do. All right, he said, you people are going to have to spend a lot of money finding out if you have oil. If I find it in this well, you'll know you've got it. If I don't find it, you can save your drilling money. What he wanted, he said, was fifteen thousand dollars "dry hole" money. If he hit, Humble didn't have to pay a cent. If he failed, Humble would contribute fifteen thousand dollars to his drilling expenses. Humble agreed to the deal.

"I was eloquent," Halbouty told Fay.

The well was deviated five hundred feet toward the dome with the hope of finding the Frio sufficiently high on the flank to contain oil. As he and Hardin reported in the *Bulletin* of the American Association of Petroleum Geologists: "This was done, and the South Boling field was discovered."

The next two wells were good, confirming production and his geology. The third one was something else. It was to be drilled to sixty-five hundred feet. Just short of that depth, however, Jack Colle, a consulting paleontologist who had been retained for micropaleontological determinations on the lease, reported that the drill had reached the Jackson shale. This was heartbreaking news. The Jackson shale was an older formation than the Frio, and thus was always below it. Oilmen quit drilling when the bit reached the Jackson shale; there was no Frio present or the bit would have reached it first.

Halbouty couldn't believe the report. He *knew* the Frio was there. He headed for the drillsite for a personal inspection. On the way he saw that a Humble well, also to be taken down to sixty-five hundred feet, was still being drilled. What he didn't know was that the Humble paleontologist also had found evidence of the Jackson shale, but couldn't find anyone with authority to stop the drilling because it was a weekend.

Halbouty arrived at his well to find that the drill had passed through the Jackson shale and, unbelievably, had penetrated a rich section of Frio! The Humble well also reached the Frio before the paleontologist found anyone in authority to shut it down!

As Halbouty and Hardin wrote in the *Bulletin* of the American Association of Petroleum Geologists:

> The data from these two wells changed the conception of the geology on this flank and made it obvious that an unusual structural condition was present. Continuous drilling operations . . . were conducted with the knowledge that this exceptional, yet unexplained, geologic condition existed. Many theories were discussed by the writers in an attempt to solve the problem, and after much study it was conclusively decided that a thrust fault of more than 1,000 feet of throw had occurred and had moved the older Jackson shales over the younger Frio beds.
>
> Although such a conclusion was contrary to the established thinking on the Coast, the facts were considered by the writers to be too strong to dispute.
>
> Never before had such a phenomenon been reported in the Texas and Louisiana Gulf Coast province. The presence of multiple, thick, highly permeable and porous oil sands under an older overthrust formation revealed a heretofore unrecognized Gulf Coast geologic condition.
>
> If such a thrust fault occurred at Boling, it is logical to assume that it occurred on other piercement-type domes, and therefore this discovery may open many thousands of so-called condemned acres on piercement-type salt dome flanks for re-exploration, and affords an entirely different slant on geological thinking in the drilling on structures of this type. . . .

It is difficult to say which gave Halbouty more pleasure— the oil or the phenomenon. Before the strange well was drilled, he and Hardin had estimated that their leases would yield about a million barrels of oil. Now they placed their estimate at four times that amount. But other geologists, paleontologists, and oil operators wanted to hear about the phenomenon. Halbouty discussed it before more than a dozen scientific societies and patiently explained it to scores of oil writers. The teacher in him had never been happier.

The South Boling discovery made Halbouty a millionaire at 42. He did not suck hind teat there as he had at South

Liberty. At South Boling, he put the deal together and remained the operator. Josey and his associates (not a gambling figure among them) were invited to participate and leaped at the opportunity. The pie was cut many ways, but Halbouty retained more than a third of the oil in the reservoir. (At May 1, 1978, the leases had yielded 3,937,194 barrels of oil and one lease was still producing.)

A nice slice of the pie went to Harry Sims, the lawyer who had shared chili lunches and dreams with Halbouty when they were living hand-to-mouth in the late thirties. Another slice went to Jack Colle, the paleontologist who knew Jackson shale bugs when he saw them even if it did appear to make the world upside down. "Gentlemen," Halbouty told the Houston Geological Society, "don't ever ignore anything unusual your paleontologist tells you. He might be nuts, but play along with him, and it may pay you dividends."

He had written that it was logical to assume that the phenomenon had occurred at other domes, but at this writing no other such thrust fault has been found in the domes along the Gulf Coast.

However, Halbouty and Hardin did find a coral band girding the Boling dome. It was five hundred feet wide and varied in thickness from three feet at the dome to three hundred feet on the outside. Halbouty told reporters that the dome was once an atoll like Bikini or Eniwetok, and the coral band was 36 million years old. He promptly drilled into it, as did Sinclair Oil, and achieved a small production.

seven

GLENN McCARTHY's Shamrock Hotel had sobered up since its momentous opening on March 17, 1949, but in late 1950 it still was a gathering place for movie stars and other notables. It was there that Halbouty first met Bennett Cerf, president of Random House, Inc., and an elfin thigh-slapper on the television show *What's My Line?*

Halbouty was sipping Scotch downstairs in the Cork Club with his old buddy James Clark, the oil writer, and Clark's strikingly beautiful wife, Estelle, when someone mentioned that Cerf was upstairs. For several years Clark and Halbouty had been collecting material for a book about the Spindletop oilfield. They had planned to have it on bookstore shelves by January 10, 1951, the old field's fiftieth anniversary. At Halbouty's goading, Clark already had written several chapters. But they needed a publisher, and Halbouty was not one to let opportunity pass.

"Go phone the son of a bitch and invite him down for a

drink," Halbouty told Clark. "We'll never get a better chance."

Clark had handled the publicity for the Shamrock's opening and had continued as McCarthy's public relations advisor. In his official capacity he sent Cerf an invitation to join them for a drink, and Cerf accepted.

The Cork Club was full of people, many of whom recognized Cerf. He could not resist a welcoming smile or an outstretched hand, so it was several minutes before he reached the Halbouty-Clark table. He found two men seated with a woman whose smile he would later describe as the sweetest he had ever seen. "At first I couldn't believe she was with them," he told a newspaper acquaintance. "Then they turned on the charm and I could understand."

He was intrigued from the moment Clark told him about some of the pioneer wildcatters and Halbouty described a salt dome. Yes, he told them, he wanted to publish the book. It would be called *Spindletop*. When Clark looked at him questioningly, Cerf wrote on a paper napkin: "I hereby guarantee that Random House, Inc., will publish a book called *Spindletop* to be written by Jim Clark and Mike Halbouty, if it is worth a darn, and a standard contract will be in the mail to them as soon as I reach New York."

The contract was not signed until a month after the Spindletop anniversary, but Clark and Halbouty had a manuscript in New York by early June. The book was released in September 1952.

It hit the bookstore shelves at the same time as *Giant*, Edna Ferber's novel about Texas and Texas oilmen. *Giant* infuriated rich and semi-rich Texans, and what amounted to a boycott limited sales of the book in the state.

Spindletop, on the other hand, was widely praised and distributed in Texas and elsewhere. Book reviewers who dealt spitefully with *Giant* lavished column inches on *Spindletop*. Several Texas newspapers serialized the book. And in Beaumont

Clark and Halbouty were greeted as heroes at several festive occasions.

Halbouty loved it all. "He would have rather co-authored that book, with his name on it, than to have found a dozen oilfields on the same day," Clark told a friend.

The book was reviewed favorably in most of the press, but one reviewer who found the book "complex" got a short note from Halbouty: "Anything is too complex for you, you son of a bitch!"

More than twenty-one thousand copies of *Spindletop* were sold. Clark would write more books, some to be published by Random House, and he and Halbouty would collaborate again on *The Last Boom*, a history of the East Texas oilfield, also published by Random House. Their enduring friendship was marked by a thousand cuss fights and as many reluctant reconciliations. It was Clark who first called Halbouty "The Great Intimidator," but he broke the nose of an oilman who contemptuously dismissed Halbouty as "Silent Mike."

(Halbouty was in California when Clark died in Houston in 1978. Halbouty immediately changed his plans. He told colleagues he would return to Houston for the funeral services, but he suggested in a roundabout way that it would be helpful to him if the services were held up for several hours. "Certainly not," said Estelle Clark. "It will tickle Jimmy pink to know he's inconvenienced Mike Halbouty one last time." Halbouty arrived on time.)

Halbouty had become a father in July 1949. He was in Louisiana, inspecting his holdings in the Pine Island field, when his secretary, Viola Vermillion, phoned. "The doctor says Mrs. Halbouty's water has broken and you're to get back here," she told him.

Halbouty argued—he *had* to argue: "Hell, the baby's not

due until the last part of September. That's what the doctor said."

'Mrs. Halbouty is in the hospital," Viola said patiently, "and the baby doesn't know what month it is."

Halbouty's stepson, Tommy Kelly, was working in the Pine Island field as a roughneck, and he flew to Houston with Halbouty. They arrived about an hour before Linda Fay Halbouty was born in St. Joseph's Hospital. She weighed less than four pounds and wasn't much larger than Halbouty's hand. She survived in the hospital premature nursery while Fay recuperated.

With the child safely at home, Dr. Herman Gardner, the obstetrician, dropped by Halbouty's den one evening. Halbouty gave him a drink. They talked of this and that, and Halbouty made them a second drink. As Dr. Gardner raised his glass to his lips, Halbouty said, "What the hell are you here for?"

Dr. Gardner lowered his glass, and said, equally bluntly, "I want some of your money." He explained that he had been named chief of obstetrics and gynecology at St. Luke's Hospital, then in its formative stage. The premature nursery at St. Joseph's Hospital, where Linda Fay had been cared for, was a good one, he said, the best in the area. "But we want to have an outstanding nursery at St. Luke's," he said. "I want to send out people to study the latest and best equipment, and I want to be able to buy it. I want you to finance the project, Mike."

Dr. Gardner knew only that Halbouty was an oilman; he presumed he was wealthy. But at that time Halbouty had income only from the Pine Island and South Liberty fields and from some consulting work. He was still living one step ahead of his income. But he told Dr. Gardner, "I'll do it." He contributed enough money to get the project underway, and continued to contribute through the years as his income increased. There were those who said Halbouty borrowed the money to get the project started. He says he doesn't remember. In any event, the Linda Fay Halbouty Premature Nursery at St. Luke's is regarded as the finest in the Southwest, and serves a much larger area.

Tommy Kelly was seventeen and a freshman at Texas A&M when Linda Fay was born. He was studying geology and working in the Pine Island field during the summer recess. He had wanted to attend the University of Oklahoma, where Fay and some of her kinsmen had matriculated, but the blustery Halbouty had used his trading guile to launch him toward A&M. "I'm afraid to tell him I want him to go to A&M," Halbouty told Fay. "He'll tell me to go to hell." So Halbouty spoke highly of the University of Oklahoma while he laid his plans.

He was scheduled to address geology and engineering students at A&M, and he invited Tommy to go along. The address was to be in the evening, but he suggested they arrive at the school in time for lunch. In the meantime he had called some of his friends on the faculty. "Now, Goddamnit, you people sell him on A&M," he told them.

Halbouty managed to slip away while the professors took Tommy to lunch and gave him a grand tour of the campus and buildings. Then they took him to Halbouty's presentation. On the way home that night, Tommy told Halbouty, "That sure is some school, isn't it?"

"It sure is," Halbouty said.

"You know, I might change my mind and go there," Tommy said.

"If you do, I think you'll like it," Halbouty said.

"I'm going to think about it," Tommy said.

He thought about it—and went to A&M. For Halbouty, it was a major triumph.

The relationship between the two had been cordial, but it had solidified the winter before Tommy chose A&M. That winter Tommy had asked Halbouty to take him duck hunting. Halbouty had never been hunting and knew little about it, but finally he succumbed to the boy's wistful blandishments.

He called a friend, Doug LaFleur, who was a professional hunting guide in Chambers County. The men had become acquainted when Halbouty drilled some wells near LaFleur's property. He told LaFleur he wanted his help in insuring that Tommy killed some ducks. LaFleur agreed.

The boy was beside himself with excitement when Halbouty bought them shotguns. Halbouty would steal his gun out of the house and practice shooting it when he went out to examine a well; he didn't want the boy to think him a tyro.

The hunt was on a cold, rainy November morning. The hunting parties were carried out to lake blinds in boats, and LaFleur managed to spend much of his time with Halbouty and Tommy. "It was a terrible morning," LaFleur later told James Clark. "Wet and cold as a witch's tit. Mike looked like he was literally freezing to death . . . like he was saying to himself, 'What in the hell am I doing out here?' He was miserable. But about that time Tommy turned to him, his eyes as big as coffee cups. I remember he said, 'Mike! Isn't this great! Isn't it!' And I'll swear to you, right then Mike forgot all about how miserable he was. He grinned and said, 'It sure is, Tommy.' And from then on he enjoyed himself as much as the boy did."

From that day on, Halbouty realized that part of his life would have been missing without Tommy—a father-son relationship. In the past, he had often been too busy to share much of his life with the boy, but the hunting trip had forged a bond that would never break.

It will be recalled that the South Boling dome discovery in late 1951 had made Halbouty a millionaire. The next few years were tumultuous ones, full of triumphs and disasters, dry holes and strikes. He had twelve rigs running. He drilled dry holes in California, in Nebraska, and in Oklahoma and Kansas. He used a barge to drill a dry hole in Lake Verret in Assumption Parish, Louisiana, and General Crude came along behind him and found an oilfield. He found oil at Saratoga in Hardin County, then frittered away most of the profit on development wells that teased him into drilling more.

But he discovered the North Lochridge field in Brazoria County, Texas, the Justina field in North Louisiana, and the Poesta Creek field in Bee County, Texas, which produced gas steadily until late 1977. And he found oil in Alligator Bayou, Texas, in an area bracketed by sixteen dry holes drilled by others. He was staying ahead of the wildcat average of one in ten.

He somehow found time to visit Europe and Mexico, become a director and later president of the Petroleum Club, and join Fay in the city's flashy social whirl. He also launched a search for uranium in Colorado—which taught him to stick to hunting oil.

George Hardin by now was Exploration and Production Manager with two geologists, John Walters and Fred Howells, under his supervision. There was a land department, an accounting department, a drilling and production department, a drafting department, and James Noel, an outstanding attorney who had left the prestigious firm of Butler, Binion, Rice and Cook to become Halbouty's general counsel.

The Shell Building was becoming too confining, and besides, Shell wanted the entire building for its own operations. Halbouty wanted to move into the Bank of the Southwest Building. Noel checked it out and reported that the necessary space would cost fifty-one thousand dollars a year. Halbouty exploded. The price was ridiculous. He'd build his own fucking building! Noel smiled agreeably.

Halbouty had spoken impetuously out of his anger, but driving home that evening he asked himself, "Why not?" He owned almost four acres far out in the county in the 5100 block of Westheimer Road. He had bought the land as an investment because he believed the city would grow in that direction eventually.

He told an architect, Wylie Vale, what he wanted—a spacious, one-story building with big offices and flower gardens. "Kind of home-like," he told Vale. "I'm never going to move."

Vale pointed out that it would be a building stuck out amidst the fields of Italian truck farmers, that Westheimer Road was a dirt lane topped with oyster shells, that there were no water or sewage lines in the area, and that Halbouty would have to dig a cess pool and drill a water well.

"Draw up the plans," Halbouty said.

He went to see W. A. (Bill) Kirkland, president of the First National Bank. He already was in debt to the bank, but he told Kirkland he wanted two hundred thousand dollars to erect a five-hundred-thousand-dollar building.

"Where are you going to build this thing?" Kirkland asked.

Halbouty told him.

Kirkland shook his head. "Not a chance, Mike. You don't know what you're doing. Nobody would ever come out there to see you."

"They'll come if they *want* to see me."

"Maybe, but I'm not going to lend you the money."

"I'll get the money," Halbouty said.

He got it—from Rice Institute, now Rice University, borrowing from the endowment fund at 4½ percent interest with a twenty-year pay period.

As the building was going up, Halbouty shot doves around his property, which he ate for breakfast. Sheriff C. V. (Buster) Kern came out to see what all the shooting was about. Halbouty told him.

"These farmers are raising hell," Kern said. "They say

you're disturbing the chickens and the cows and I guess the Goddamned sheep, if they've got any."

"It's my property and my doves," Halbouty said, "and it's not against the law out here in the county."

Kern agreed. "But I don't think you ought to do it, law or no law. You're the only business around here, and maybe you ought to be making friends instead of enemies."

"By God, you're right!" Halbouty said—and he put away his shotgun.

He went to the Rio Grande Valley and brought back thirty-three palm trees for which he paid three hundred dollars each. A freeze killed them. He replaced them and a freeze killed the replacements. He said to hell with it and planted crepe myrtles.

But a wiry, feisty fellow with dirty fingernails and ankles made Halbouty's landscape a showplace. His name was Hugh Russell. He introduced himself to Halbouty one morning as the building was being completed. He was the world's foremost grower of day lilies, he said, with 3½ million of them blooming at his gardens in the northern part of the county. He had checked Halbouty's soil and had found it perfect for the *hemerocallis*.

He had a proposal. He would plant Halbouty's garden area if Halbouty would leave it open for public viewing. If he planted now, he said, the blooms would be ready for inspection by visitors to the International Hemerocallis Society convention, to be held in Houston. There would be no cost to Halbouty, and it would be the most beautiful garden in Texas.

"Go ahead," Halbouty said. "If it's as pretty as you say it will be, I'll pay your costs."

The bulbs were planted and the lilies bloomed. The convention visitors came and went. A few days later Halbouty saw Russell and his son, Jake, examining some of the lilies. He invited them into the new building for coffee. "They're prettier than you said they would be," Halbouty told Russell. "What was your cost?"

"With labor and all, about five thousand," Russell said. Halbouty wrote him a check for ten thousand dollars.

In the final stages of the building's construction Halbouty began one of the enduring relationships of his life. In the crew that installed the air-conditioning system was a black man in his forties, William Prince. Halbouty, nosing around every corner of the rising structure, became aware of Prince's competence. For his part, Prince was fascinated by Halbouty's combustible style.

Shortly after the building was open for business, Prince came to see Halbouty about a job maintaining the air-conditioning system.

Halbouty countered, "I'd like you to be my building engineer. Take care of the whole damned building. What do you say?"

Prince took the job, and gradually began assuming responsibility for a number of other jobs. A couple of weeks later Halbouty saw Prince working in the flower garden. Not long after that he sent Prince to the airport to pick up a business associate. In a short period, Prince made several such trips, and when Halbouty summoned him the next time, Prince was wearing a chauffeur's uniform.

"For Christ's sake, Bill, what are you doing in that getup?" Halbouty demanded.

"I'm doing it right," Prince said.

When Prince saw that Halbouty was beginning to entertain investors in his offices by using a catering service, he got a batch of women's magazines and studied table-setting, serving, cooking, and bartending. Then he took over those chores.

"Let Bill do it," Halbouty would say when some new chore emerged. "He'll get it done, and he'll do it right."

Thus Prince took charge of the mailroom and the copying department.

Prince was the only black in Halbouty's employ, but apparently he was not a token. Halbouty treated him just like he treated his other employees—tyrannically one day, considerately the next.

But Prince's jobs put him in close contact with Halbouty, and over the years a friendship was forged. Neither man made much of it. "We look out for each other," Halbouty told a visitor one day.

Houston was on its way to becoming a cosmopolis, and with astonishing rapidity it grew to Halbouty, then passed him, then surrounded him, then stretched toward the western horizon. The shell road became a thoroughfare. And just east of Halbouty's building, also fronting on Westheimer, there erupted a glittering mall of high-fashion stores, hotels, office buildings, and specialty shops called the Galleria.

Its originator was one of the entrepreneurs who appear regularly on the Houston scene to inject another shot of high life into the already fractious city. His name was Gerald Hines, and he came to Halbouty to make a deal. He wanted Halbouty's land, for his master plan called for him to move westward. Halbouty told him he was content where he was.

"I'll surround you with buildings," Hines said.

"Go ahead," said Halbouty.

Again and again Hines came back, raising the price he would pay each time. Another bidder got in the play, offering Halbouty more than two million dollars. Halbouty still said no. "I love this place, and I don't need the money," he maintained

He was working late on a cold Christmas Eve, alone in the building, when he heard a tapping at his office window. Hines had seen his light and wanted in. "You scared my ass off," Halbouty said when he opened the door.

Hines said he had an offer Halbouty couldn't refuse. "I'll give you a percentage of everything in the Galleria, what's there now and what I'll build. Every time somebody buys a

Coke, you'll get a percentage of it." He said a Lord & Taylor's and another hotel would edge onto Halbouty's space.

Halbouty told him he'd think about it. What Hines had said had made him think of his daughter, Linda Fay. A deal such as Hines had offered could assure her—and the grandchildren he hoped would come—of a rich future far beyond Halbouty's lifetime.

Hines, meanwhile, had obtained land across Westheimer from Halbouty. It was not part of the original offer, but he said he would make it available if Halbouty wanted to build there.

Halbouty took Hines's deal, and across the street he erected a five-story office building, which was named The Halbouty Center. It was constructed before his old building was razed. He was standing at a window in his fifth-floor office when the first bulldozer moved in to level his home-like building. When the first brick fell, he wheeled away from the window, tears in his eyes.

"Mary!" he called out to his new secretary, Mary Stewart. She came into his office. "Yes, Mr. Halbouty."

"I'm going to California! Call me when they finish tearing that son of a bitch down!"

In the mid-fifties he was invited to present a paper at the World Petroleum Congress in Rome because he was the recognized expert on salt dome geology. Fay went with him. He and George Hardin combined their knowledge and vision in the paper to fit Halbouty's eloquence.

The leader of the U.S. delegation was Ernest O. Thompson, chairman of the Texas Railroad Commission, the state oil and gas regulatory body. At a party the night before the meeting, he called Halbouty aside. He had read Halbouty's paper.

"The Russians weren't invited, Mike," he said, "but

they're going to have some unofficial scientists here. Much of your stuff will be new to them. And they're sure to ask you a lot of questions. Perhaps you ought to cut your paper some. And you can be vague in your answers in the questioning period."

The Cold War was frigid at the time.

"I've learned that they're more interested in your paper than in any other," Thompson said. "I suggest that you do as I say in the national interest."

Halbouty told him he couldn't agree completely. "I won't tell them about the newest tools for salt dome exploration and drilling. I won't tell them *how* to do anything. But I'm going to tell everybody about salt domes. I mean, all earth scientists . . . and I don't give a damn where they come from." It shocked him to think that scientists should withhold basic information from one another.

"But, Mike . . . " Thompson said.

"I know I'm a right-wing son of a bitch," Halbouty said, "and I always will be. But don't tell me what to do about basic information with other scientists."

He delivered his speech and answered questions from the floor he thought proper. He edited his speech, and his answers to questions, according to his own guidelines.

eight

BEFORE WORLD WAR II, gas was important primarily as a reservoir energy source. "Free gas" rested like a cap on the crude; "gas in solution," of course, was gas dissolved in the crude. Both forced the oil to the point of lowest pressure, the borehole. Once it had lifted the oil from the reservoir, gas had served its purpose. It was "flared"—and an airplane pilot could fly from Houston to New Orleans at night with no other beacons than the burning gas.

A wildcat well that found only gas almost always was capped and abandoned. Some of today's great gas fields were so ignobly treated, and the gas remained buried until the industrial explosion created by the war. Even then it was not just the gas itself but a means of transporting it that unlocked the door to its future.

As a wartime measure, the U.S. government constructed two great pipelines from Texas to the refining complexes in the New York and Philadelphia areas. They were called the

"Big Inch" and the "Little Inch," the former being twenty-four inches in diameter, the latter twenty inches. The "Big Inch" was the largest oil pipeline ever laid.

In June 1945, with victory in Europe in hand, *World Petroleum Magazine* summed up the pipelines' contribution to the war effort: "Allied invasion of enemy territory in Europe would not have been possible, nor could it have been sustained, without the aid of the 'Big Inch' and 'Little Inch' lines. Constructed in record time, under wartime limitations, the two lines have delivered to the Eastern Seaboard over 316,000,000 barrels of crude oil and refined products."

But the fighting had hardly stopped when a struggle began for control of the two pipelines, which the government no longer needed. It was obvious that the great lines, with feeder lines sprouting out like ribs from a backbone, were the perfect conduits for Texas gas to the gas-hungry East. A newly formed organization, Texas Eastern Transmission Corporation, won the prizes from established companies, and the fruit of peace began to flow eastward. Other great lines were laid in other directions, even to California, where Texas gas provided heat and electric power from San Francisco to Los Angeles.

Gas powered the war boom and the postwar boom. It was cleaner than coal and fuel oil when burned in the spreading plants and factories, and more efficient as a fuel.

So oilmen who had spurned gas now sought it. Among them was Mike Halbouty, and he found it.

In 1956, Halbouty and his geologists drilled fourteen exploratory wells and twelve of them found production. This almost incredible achievement was due as much to his aggressiveness, his willingness to drill where others feared to risk their capital and reputations, as to his brilliance and the brilliance of his geologists.

Some of the fields found were losers, some were marginal, some were commercial, and two were humdingers.

One of the humdingers was the Pheasant field in Matagorda County. Halbouty began the play there by obtaining a 160-acre lease in an area that had never felt the drill but where George Hardin and John Walters had mapped an anticlinal structure. Since this was not a salt dome where a gambler might find a fortune beneath a postage stamp, Halbouty wanted all the acreage he could gather.

Sorry, said his land department. Your 160 acres are surrounded by leases controlled by majors and large independents.

It was not a unique situation. Companies always had been loath to drill on small leases of 400 to 500 acres, and they always had been reluctant to join such leases to form a proper drilling block that they could share. In many oil areas, potential fields were left undrilled for years because they lay beneath several small leases owned by several companies. "Sit and wait" seemed to be the philosophy behind the inaction; "Sit and wait and let's see what happens."

Halbouty was the catalyst. He went from company to company, showing his geology. Give me a percentage of your acreage, or give me dry hole money, he would say, and I'll drill for all of us. One by one they fell into line. So certain was he of success that he spread out beyond the majors, leasing any land he believed to be within the limits of the prospect. His interest in some sections of the block was 10 percent, 20 percent in others, 100 percent in still others.

The first two wells found two rich reservoirs holding gas condensate.

Some gas fields yield what is known as "dry gas," so called because it contains no liquids. Most of those on the Gulf Coast yield a "wet gas," whose heavier molecules become liquid at the surface. The liquid is called *condensate* or *distillate*. It is separated from the dryer gas at the surface and pumped into tanks. The dryer gas is piped to a gas plant where any remaining liquids are removed.

Condensate brought about the same price as crude because refiners had learned it could be converted to gasoline and other "thin" products more cheaply than crude. And there was a plus for the oilman: he could produce condensate to the limit of a well's daily gas allowable, and generally that far exceeded the amount of oil he was permitted to produce daily from an oil well. Thus, he was assured of a quicker payout and quicker profits.

So Pheasant field was a rich find. By May 1978 wells on Halbouty's leases had yielded 46,384,870 mcf (thousand cubic feet) of gas and 932,285 barrels of condensate—and even 19,112 barrels of oil. And most of the wells were still going strong.

The second fine strike of 1956 was in an old oil patch he had long promised himself to revisit, the West Hackberry field in South Louisiana. It will be recalled that Yount-Lee had drilled there in the early 1930s and had found two hundred feet of oil sand. However, the well blew out and was lost. Miles Yount died and the company had no impetus to seek the sand with another well. When Yount-Lee sold out to Stanolind Oil & Gas, the outside appraiser, D'Arcy Cashin, had ignored the two hundred feet of oil sand in evaluating the West Hackberry leases—and this was one of the reasons that the young Halbouty had challenged Cashin's judgment in that stormy trading session.

Stanolind, by now Pan American Petroleum Corporation (it later would become Amoco), had found oil on the leases after the trade. Others also had made strikes around the salt dome. But now, more than twenty years later, Halbouty and Hardin studied the area and concluded that plenty of oil remained to be found on the southeast flank. Halbouty went about putting together a block, picking up leases and farmouts.

The first well found a prolific new reservoir in the Camerina sand. By May 1978, Halbouty's West Hackberry leases had produced 3,169,029 barrels of oil and were still productive. His 1978 income from these leases alone was about $40,000 a month.

Halbouty had only a single partner in the 1956 drilling program, Empire Bank & Trust Company of New York. Such an arrangement, when it could be negotiated, was a blessing to an oilman if for no other reason than that it simplified his bookkeeping and reduced his areas of accountability. Empire paid 100 percent of the cost of the first well in a new area where Halbouty had explored and acquired leases at his own expense. After discovery, costs of other wells were split evenly; profits were divided fifty-fifty after payout. Both parties, for example, put up about three million dollars for the 1956 exploration program. Empire Trust had the option of non-participation in the drilling of any new area.

Empire Trust also was Halbouty's only partner in his 1957 exploration program—and 1957 was another banner year. Ten exploratory wells were drilled. Eight were dusters, but two were significant discoveries. One discovery would thrust Halbouty into the fiercest fight of his professional life. It also would bring him a deep satisfaction, which he would never publicly admit.

Halbouty was drawn to the Fostoria area, some forty miles northwest of Houston in Montgomery County, by two dry holes drilled by a major oil company. What intrigued him was the "flatness" of the formations. The dry holes were six miles apart, yet the formations were the same. That is, a shale formation at six thousand feet in the first well was at six thousand feet in the second. There was no variation in the depth and thickness of the formations.

Six miles, Halbouty thought, was a lot of distance without

a fault appearing. Somewhere across those miles a structure could be buried. He and Hardin and John Walters studied the area and found what they considered indications of a structure. A geophysical examination confirmed its presence.

The area had been ignored after the two dry holes, and not just because of them. Oilmen believed that even if hydrocarbons were present in the area they would be in the Wilcox sand, a "tight" sand that yielded its wealth reluctantly. This didn't disturb Halbouty. He began leasing.

At one point in the leasing he ran into a snag. Some acreage he deemed necessary to his block was under lease to a Houston drilling contractor, Cecil Housh. Housh hadn't drilled, showed no intention of drilling, and he wouldn't deal with Halbouty on any terms Halbouty offered. Hardin finally sent the land department to "top-lease" Housh. The landmen dealt with the acreage owner, leasing the acreage from him with the lease beginning the minute after the Housh lease expired.

The Housh lease had six months of life. Thus Housh had to drill within six months or he would lose his lease. If he did drill, Halbouty would lose his lease money, but it would be worth it; the Housh well would prove whether the acreage Halbouty already had accumulated was productive or barren. If Housh didn't drill, Halbouty would have the acreage under lease in six months.

"I'm looking down your Goddamned throat," Halbouty said when an angry Housh stormed into his office. "You say it's a dirty deal because we top-leased you. We want to drill a well and you won't move, you won't participate in any way. What the hell do you expect me to do?"

Top-leasing was not an uncommon practice, but it was a power play. Housh was not in a position to drill, and could not be ready for a year. He accepted a deal that gave him an interest in possible production, and it made him rich. For at

10,880 feet Halbouty found a productive Wilcox sand, and two other wells found two additional productive zones. Again he had gone in where others feared to tread. This time he controlled all of the wells in the field.

And Fostoria was a winner. By May of 1978 it had produced 45,798,314 mcf of gas and 1,042,315 barrels of condensate, and was still productive.

During World War II Glenn McCarthy went into the Port Acres area immediately south of the city of Port Arthur in Jefferson County. He was hunting gas, for which the wartime industries had created a vast market.

He drilled several wells and found little or nothing. He capped his final duster at 10,585 feet.

Had he drilled just 19 feet deeper he would have tapped one of the richest reservoirs on the Gulf Coast!

Thirteen years later, in 1957, Mike Halbouty drilled that additional 19 feet, and his well was only 1,500 feet from McCarthy's final duster.

Halbouty went to Port Acres when he learned that two leases were available on either side of a lease where Byron Meredith, another wildcatter, was preparing to drill. A geophysical crew had located what was considered a small structure.

Meredith had 640 acres under lease. Hardin went to see Meredith's geologist, "Red" Elwood. He told him Halbouty was about to acquire the 640-acre leases sandwiching the Meredith lease. Did Meredith want to join Halbouty in buying them?

No, said Elwood. The structure was small and it lay entirely beneath the acreage Meredith was getting ready to drill. Meredith had all the land he wanted. But, said Elwood, if you think differently, why don't you get the two leases and give us some dry hole money? If we hit, you owe us nothing.

If we fail, you'll know your leases are dry and you pay part of our drilling costs.

Halbouty liked the idea. He made the dry hole deal on a per foot basis; he would pay a percentage of the drilling cost of each foot drilled. He acquired the two leases for $19,200, $15 per acre.

Meredith began drilling. His bit bored into the structure at 9,100 feet. He had found nothing liquid or gaseous. He was through.

Halbouty was confounded. He had been certain that Meredith would drill deeper, and had been prepared to pay his share of the cost. He knew that six miles to the southwest a thick but barren bed of Hackberry sand had been found below 10,000 feet in a dry hole drilled some years earlier. He had expected Meredith to drill at least that deep.

Someone else had been watching Meredith with great interest, a slender, intense young geologist named Tom Barber, assistant divisional exploration superintendent for Pan American. Pan American owned acreage adjoining Meredith's lease on the south. Barber also knew about the dry Hackberry sand in the old well to the southwest.

When Meredith stopped drilling, Barber moved quickly. He gave Meredith $60,000 to deepen his well, and Meredith gave Pan American a half interest in his lease. Barber told Halbouty about the transaction. Halbouty agreed it was a good deal for both parties. He and Barber were aware, however, that it still was a big gamble.

Halbouty set out to acquire more leases. He found that Pan American was leasing also. The prices for leases were going up. Halbouty went to Pan American officials. "These land owners are going to take our shoes if we don't stop bidding against each other," he said. "Let's get together." The officials agreed, and Halbouty and Pan American began a joint leasing program.

Empire Trust had backed off from the Port Acres venture

when it was first broached and while Halbouty was acquiring the two leases sandwiching Meredith's. Some of their petroleum engineers thought Halbouty was too high-handed, that he seldom took notice of their opinions. Despite the money Halbouty had made Empire Trust, the officials listened to their professional people.

With considerably more acreage under lease, Halbouty called a meeting of his staff. He had been elated when Empire Trust had backed off the Port Acres play, he said, but things had changed. "Since I've picked up this additional acreage, they can say I didn't show them all I had, that I didn't give them the full picture," Halbouty explained. "Another thing, they went through the bad with the good with us, and I think it's only fair to give them a chance to get back in."

"You have no legal obligation to them at all, Mike," said Noel, the attorney.

"And no moral obligation," said Hardin. "They pulled out of their own free will."

Other staff members supported Noel and Hardin. Staff consensus: To hell with Empire Trust. Halbouty finally called a halt to what was becoming a stormy meeting. He had made up his mind, he said. He was going to New York and open the door to Empire Trust.

He flew to New York and met with Empire Trust officials. He explained his position; he invited them back in the deal. They thanked him for coming; they appreciated his attitude.

"We'll let you know in several days what we intend to do, Mike," the chairman said.

"No, you'll let me know now," Halbouty said. "I'm not going to leave here without an answer. I'll step outside five minutes for you to talk it over, but that's all."

"Five minutes, then," the chairman said.

Ten minutes later Halbouty was on his way back to Texas.

Empire Trust had accepted his offer to enter the Port Acres play on the old terms.

Meanwhile, Meredith had resumed drilling. At 10,534 feet the well blew out. No deep casing had been set and the drilling mud was too light. Gas screamed out of the drill pipe. Myron Kinley, the famed wild-well-fighter, rushed to the drill site.

There was no hope of pulling the drill pipe from the hole and setting casing. "We're going to have to use that drill pipe for casing," Kinley finally said. "We'll cement it in, and then all I have to do is figure out a way to cap it."

It took four days to cap the well, and the strange device Kinley employed was so tall that the final wellhead connection to a pipeline was thirty feet in the air.

Even before the well was capped, Halbouty was drilling on one of his adjoining leases. He figured the Meredith well, while prolific, was producing from a stringer off the main body of the Hackberry. The main body would be deeper still.

Below 10,000 feet he set heavy casing, and he used extra heavy mud to hold down the anticipated gas pressure. At 10,604 feet he found a magnificent gas and condensate sand.

Meredith had thought the structure so small that it would support only a few wells on his lease. Halbouty and Hardin had calculated that it was large enough to support at least one good well on each of Halbouty's sandwiching leases. But as Meredith's well was being drilled, their study of it led them to believe the structure was more than it had appeared to be at first. And Halbouty's well and wells that followed it showed them they had drilled into a stratigraphic trap of considerable size.

They were to find out just how large it was from a young promoter from Massachusetts. He was Peter Henderson, a successful stockbroker who had left Wall Street for a roughneck's job in Louisiana because he wanted to learn the oil business. He had worked his way up to driller, and had tried his hand at promoting in Oklahoma and West Texas.

Henderson approached Halbouty and Hardin and told them he had assembled a number of town lots in the Port Acres community itself. Railroad Commission orders called for one well per 160 acres in the Port Acres field. Henderson had made a 160-acre unit of his town lots. He jauntily explained that he planned to drill.

The Railroad Commission had been established in 1891 to police the iniquity-ridden railroad industry in the state. When the automotive age produced the bus and trucking industries, the legislature placed them under Commission control. Later the state began to regulate the oil and gas pipeline systems. Because, like railroads, pipelines were common carriers, they too were placed under Commission regulation. Then, in 1919, with boomtowns sprouting up all over Texas and oil wells running wild, legislation was enacted giving the state more effective control over the conservation of its natural resources. Because of its regulatory experience, the Commission was assigned the task of enforcing the new law. From that point on it grew like a banyan tree until its oil and gas division regulated almost every aspect of the state's oil production.

Henderson drilled his tract and his bit found the same thick sand that was in Halbouty's wells and those of Pan American.

Immediately there was a rush to lease town lots. Promoters and oilmen flocked into Port Acres, trying to lease every lot in sight, whether it was vacant or held a house or building. Some of the lots were as small as two-tenths of an acre.

Those who picked up leases went to the Commission for "exceptions" to Rule 37, which prohibited the drilling of a well within 300 feet of a completed well or one being drilled, and within 350 feet of a property line.

The exceptions were granted in every case.

When the Commission had fixed 160 acres as the size of producing units in the field—one well per 160 acres—it had established a gas allocation formula of "one-third well and two-thirds acres." This meant that one-third of the total gas production from the reservoir would be divided equally among all of the wells in the field; two-thirds of the total production would be allocated to the wells in the proportion that each well's surface acreage bore to the total surface acreage in the field. The total gas allowable for each well, then, was the sum of its well allowable and its acreage allowable.

With the granting of the exceptions to Rule 37, the Commission was saying that a well could be drilled on each tiny town lot lease. Thus Halbouty, Meredith, Pan American, and others now in the field would face terrible losses under the one-third–two-thirds formula; 5 percent of the field—the town lots—would be producing 30 to 40 percent of the gas and condensate.

And there was another worry: too many wells would mean a drastic reduction in the pressure necessary to push the condensate to the surface; millions of barrels of condensate would be irretrievably lost.

Before any of the town lots could be drilled, Henderson asked the Commission for an amendment to the allocation formula. Halbouty, Meredith, and Pan American supported the application, and also asked that the Commission consider whether waste would occur under the present formula, and whether recycling of gas to promote better recovery of the condensate should be instituted and conducted. They had in mind the construction of a plant that would send gas back into the formation to maintain reservoir pressure and lift the condensate to the surface.

The Commission refused to change the formula, and it refused to require recycling or pressure maintenance in the field. And it refused to halt the drilling on the town lots.

Granting of exceptions to Rule 37, the spacing rule, was a Commission practice of long standing and had been supported time after time in the courts, the Commission said. The exceptions were designed to protect the vested right of an owner to produce substantially the oil and gas underlying his property. To deny this right would amount to confiscation of his property.

The Commission asserted it had no authority to order a recycling program, nor could it order the field to be unitized—a practice whereby a few wells could produce the entire reservoir for all leaseholders with all sharing in the production. That was a voluntary matter.

Halbouty's angry reaction was published in the trade press and in newspapers around the oil states.

> This indiscriminate granting of exceptions to the applicable spacing rule is for the alleged purpose of preventing the confiscation of property of the owners of tiny tracts. It results, however, in owners of tracts as small as one-third of an acre being permitted to produce one-third of the amount of gas permissibly produced on a full-size 160-acre spacing unit.
>
> Three wells on three such small tracts containing a total of one acre are permitted to produce from the reservoir as much as one well located on a full-size 160-acre unit. These three wells would therefore be entitled to produce 160 times as much gas as was originally in place under the leases.
>
> Can anyone contend that such an allocation formula is in accordance with the conservation statutes of this state, which command the Railroad Commission to protect correlative rights, conserve gas, and allocate gas production from a reservoir each month so that each well is permitted to recover its full share of the gas?
>
> Why should a man with a half-acre of land be given far more than his share of hydrocarbons below that land? What kind of American justice is that?
>
> Some people are saying we are trying to "get the little man." What we are trying to do is to deny him the legal right to something that by moral law, at least, belongs to some other land owner. The Commission is mistaken if it believes it is bound by laws or court decisions to grant disproportionate allowables to wells drilled on small tracts.

The continuation of the one-third–two-thirds formula, he said, was "political conservation masquerading as protection for the little man." This was perhaps his harshest statement.

And he promptly sued the Commission. As one columnist wryly observed, it was somewhat akin to an accused burglar giving the judge a hotfoot while on trial for his liberty.

Three men sat on the Commission, elected by the people to staggered six-year terms. Both knavery and rank ineptitude had blemished the Commission from time to time since its beginning, but it had remained the most powerful regulatory body in the state. Similar bodies in other oil states followed its examples. The members were courted by the independents with an ardor ordinarily found in first love affairs. The majors romanced them, but with a cooler passion. State courts had grown to lean on their expertise. Halbouty, by suing the Commission, was in a very tough fight with a very short stick.

That he sued was a measure of his indignation, for he had long praised the Commission for its conservation policies, and spoke of it in terms approaching reverence.

Owners of town lot leases joined the Commission as defendants in the suit. Meredith, Pan American, and Peter Henderson joined Halbouty as plaintiffs.

The Court upheld the Commission. Out of all the oratory, the bales of paper, and the inevitable sidebar issues that crop up in a civil suit, the decision basically said that the Commission could not deprive an owner of his vested right to recover the oil and gas under his land.

Halbouty appealed to the Third Court of Civil Appeals. He appeared to be fighting a losing battle outside the courtrooms as well as in; by now twenty-two wells had been drilled and were producing on town lots that totalled 21.3 surface acres . . . and there were five hundred town lots.

Halbouty set about leasing town lots himself, but he joined with others to form them into 160-acre producing units. He would not drill on a single town lot or a combination containing less than 160 acres.

After his initial outburst, Halbouty had managed to keep his mouth shut, as his attorneys had advised him. But he was boiling inside. He felt that if the law recognized the right of an owner to recover the oil and gas under his land, it should apply to *all* owners, not just town-lot operators. But he was stymied. To protect himself from drainage he would have had to drill to the same density as the twenty-two wells on the 21.3 acres, a practical impossibility. And the Commission wouldn't allow it, anyway. Such unlimited drilling would be in the teeth of the Commission's ruling that the drilling of only one well on 160 acres was necessary for the prevention of waste.

And his attorneys had pointed out another dismal fact: A formula he had proposed to replace the one-third–two-thirds formula could also damage him financially. His formula accorded importance both to surface acreage and pay sand thickness. He thought it was fair and based on sound geoscientific principles. But he had been aware that it could damage him since the day he first entreated the Commission.

He broke his silence at a joint meeting of the Houston Chapter of the American Petroleum Institute and the Gulf Coast Section of the Society of Petroleum Engineers. He was chief speaker of the event in the Houston club. He had been chosen to speak before the Port Acres dispute developed.

His six hundred listeners settled back for the usual rah rah speech: the federal government was a menace; incentives (the industry's euphemism for money) must be increased; the consuming public didn't understand the depletion allowance; industry critics were merely headline-hunters. Since Edwin Drake drilled the first American oil well in 1859, oilmen have absorbed these teachings with their mothers' milk. Like religious fundamentalists, they never weary of hearing them repeated, and industry speakers seldom disappoint them.

What the six hundred heard was something else—a long, well-reasoned attack on one of their institutions, the Railroad Commission; a plea for total conservation, not partial; and a five-part program to modernize the rules and regulations governing oilfield practices. As a sop, perhaps, Halbouty also trotted out the specter of federal control of the industry.

What had happened at Port Acres had prompted him to make an examination of rules and regulations at other fields in the state and in other oil states as well. So he was prepared to play the role he loved best—an articulate teacher who knew what the hell he was talking about. His ox was being gored at Port Acres, but his personal anger had been submerged in a wider, stronger passion—his sincere belief that a personal God had placed oil and gas in the earth specifically for human use. "Remember," Miles Frank Yount had told him in his youth, "some day we'll regret every drop we've wasted." To Yount, the pragmatist, waste was stupid. To Halbouty, born in the faith, waste was a mortal sin. And waste he had seen in his studies.

It may have seemed a strange time to talk about "total conservation." The oil industry was in one of its periodic "glut" phases. In 1945, the industry had expected a drop in demand at the end of the war, but demand grew. By 1950 there was a glut, but the Korean War drained it off. By 1956 there was another glut, but Egypt seized the Suez Canal and U.S. oil demand grew to fill Europe's needs.

By 1958 demand was down again, and no more crises were in sight. Wells in Texas were producing only nine days a month because the majors were bringing in so much foreign oil. Texas crude sold for about $3.50 a barrel; Middle East crude could be laid down at the Eastern Seaboard refineries for less than $1.85 a barrel.

The Texas Railroad Commission and its counterparts set well allowables and the number of days of production, but the majors ran the show. They told the various commissions how much oil their refineries would need each month, and the well allowables and number of production days were set accordingly. In the late 1950s, then, Texas' innards were throbbing with unproduced oil, and waste should have been at a minimum.

But Halbouty's passion had combined with his knowledge and his intellect to produce a vision. ("A clairvoyant," a columnist would write of him in the 1970s.) He could see beyond the glut to a time rapidly approaching when America's dependence on foreign oil would threaten her economy and her security and tax the spirit of her people. He touched only lightly on this theme in his speech; he would return to it later when he had studied it thoroughly. On this day in 1959 he preached on what he thought his audience could visualize.

At one point he said:

> Within a very few years twice as much oil will be needed to meet the requirements of this country as it takes today, despite accomplishments in nuclear energy, fuel cells, solar energy, and other energy sources. Therefore, petroleum conservation is far more vital today than in the 1930s when most of the present laws on the subject were adopted. Conservation is more important to the public interest and public welfare than it is to the petroleum industry. . . .
>
> Our domestic petroleum industry is of utmost importance to the security of our nation. The public should know that today petroleum conservation laws, rules, regulations and applications in Texas, the most experienced state on the subject, are behind conservation needs. Our citizens should be made to realize that our country needs total conservation, not just partial conservation. . . .

He went on to explain the situation at Port Acres and other fields. He described some of the opponents of total conservation.

> These people are not wildcatters. They do not discover oil or gas fields. They come in after discovery, and through wells closely

drilled under the exception to Rule 37 and produced under in-equitable allocations formulas, profit by draining their neighbors' oil and gas. . . .

There was approximately 110 feet of pay sand under the town lots, he said. This meant a one-acre plot had beneath it about 275 million cubic feet of gas and nine thousand barrels of condensate, which at current prices was worth fifty-three thousand dollars. It cost about three hundred thousand dollars to drill a well to the pay sand. Why would an operator drill a three-hundred-thousand-dollar well to recover fifty-three thousand dollars in gas and condensate? "He does it because the allocation formula permits him tremendous net uncompensated drainage from adjoining tracts. Based on the one-third–two-thirds formula, he will recover approximately twenty times more than the amount under the tract. . . ."

The unnecessary wells prematurely reduced the pressure in a reservoir and physical waste resulted. Even with that occur-ring, the Commission would not order a recycling program. "As matters now stand," he said, "there will be no pressure maintenance program or recycling program. As if the deliber-ate waste of irreplaceable hydrocarbons were not enough, a plant will not be built, jobs will not be created, property values will not increase, and overall progress of a community will suffer. For all of this, the taxpayer must pay. . . ."

His audience never before had heard the Commission thus flayed in public.

"The answer to all the critics of the oil industry is to elimi-nate the weakness of present regulations. They have served Texas well in the past, but changes now are needed. To ig-nore this fact and remain silent is to act irresponsibly."

For the backward in his audience he defined "unitization" as producing an entire field as a unit. "Pooling" was the combining of smaller surface tracts into a larger tract equal to or approximating the acreage pattern or proration unit area set by the regulatory body. (Peter Henderson properly had pooled his town lots into a 160-acre spacing unit and had drilled one well on it, and had put other town lots into a pool with other town lot leaseholders.)

Then Halbouty listed his recommendations. First, he said, "a law should be passed to make pooling on the established spacing pattern compulsory." With compulsory pooling, voluntary field-wide unitization ultimately could follow.

Second, "the use of arbitrarily selected allocation formulas should be discontinued." Each field should have a formula which would give each well and tract its fair share of the underlying oil or gas, but no more. "Such formulas should be in accord with known technological and engineering data for each particular field."

Third, "immediately after completion of the discovery well in a new field and before offsets are commenced, temporary spacing rules should be adopted for the particular producing formations encountered until the field is sufficiently developed to determine the *best* spacing for the field—and in general, wider spacing should be adopted for all fields."

Fourth, "any town or city in which, or near which, oil or gas is discovered, should immediately adopt an ordinance to protect the lives and property of its citizens from the hazards which accompany the exploration for and production of oil and gas."

Fifth, "this industry and the citizens of petroleum states must bolster conservation laws—modernize them to meet existing problems—before we are faced with federal laws that would lead directly to federal control of our industry."

There was more, but Halbouty's advocacy of compulsory pooling was the shocker. To oilmen, "compulsory" was a socialistic word that evoked the image of "public utility status" for the industry.

There was a long moment of silence when he was through speaking, and he later confessed to his secretary that his skin grew cold. But then there came a thunderous burst of applause, a standing ovation. His dark eyes had flashed during his speech; now they filled with tears. As James Clark would have said, and maybe did, it was more thrilling than finding an oilfield!

He had uttered publicly the private thoughts of many of them, had used words they most often felt should remain unsaid. As he had done in seeking oil and gas, he had gone in where others feared to tread.

As he was leaving the Houston Club he was halted by Billy G. Thompson, oil editor of the Houston *Post*. "You ought to take this fight to the people, Mike," Thompson said. "Judges and commissioners listen to the ones who elect them."

Halbouty needed nothing more. He already was on fire. And nothing else had worked. To hell with the lawyers; let them keep working in the courts. He'd tell the people what was going on.

He had bought an airplane and hired a pilot after the South Boling discovery. So he left the lawsuit to the lawyers and the business to Hardin, who was perfectly capable of handling it, and set out on what soon was labeled a crusade by the more romantic oil writers.

And it was a crusade. It *was* more thrilling than finding an oilfield.

He believed every word he said. And he believed it imperative that he be heeded, for he had looked into the future and found it grim.

nine

Halbouty had never commented publicly on his hitting the Port Acres pay sand just nineteen feet below Glenn McCarthy's wildcat failure, but several oil writers had taken note of it. When McCarthy was asked to comment, his remarks amounted to no more than a philosophic shrug of the shoulders. He had found millions of barrels of oil by going a little deeper than the herd, and his nature was such that he felt it would happen again.

Halbouty referred to it, in one sentence, in a scientific paper he wrote on the field. At the end of the paper, under conclusions, he first listed three dealing with the field's geological characteristics. Number four said: "No study of Port Acres can be complete without wondering how many other wildcats have been plugged just twenty feet above 500 billion cubic feet of gas."

It was speculation having no place in a scientific paper, and it won him no friends. In his crusade he needed all the friends he could get. Even before he began it, defenders of the status'quo were organizing to oppose him. Among those who did not oppose him, even those who had applauded his maiden speech, there were few who rushed to aid him. Ironically, the dedicated right-winger was taking to the streets to tilt his lance against the establishment. And in the beginning he virtually stood alone.

He made speeches to oil groups, but he also spoke to PTAs, Lions, Rotarians, Kiwanians, Elks, Chambers of Commerce, and any other groups he could get to sit still for his message. "He'd talk to a gaggle of geese," one reporter said, "if they'd only waggle their wings at him." Another wrote, "He is making more whistle-stops than Harry Truman."

And the Third Court of Civil Appeals upheld the lower court. Halbouty immediately took his case to the State Supreme Court.

He began gaining support. Other oilmen let it be known publicly that they agreed with the thrust of his arguments. Opponents called him a "tool of the majors." Halbouty told reporters: "You tell those bastards that the majors know I'm not anybody's tool. But if majors happen to agree with me, then I'm glad to have them on my side." Reporters were always happy to clean up his quotes.

The charge had been prompted by remarks made at the American Petroleum Institute meeting in Chicago. M. J. Rathbone, chairman of Standard Oil Company of New Jersey, said that waste from needless wells was crippling the industry. And R. M. Williams, counsel for Phillips Petroleum Company, practically echoed Halbouty's speeches. He said the exceptions to Rule 37 "can best be characterized as the privilege of piracy." The phrase, "privilege of piracy," was one Halbouty must have wished that he had thought of first.

Opponents pointed out that Halbouty had drilled on many small tracts in his time. It was a low blow. The small tracts he had drilled were on salt domes where there were no spacing rules because of a dome's geological makeup. On a salt dome, as Miles Frank Yount had told the young Halbouty, every well was a wildcat.

Months passed and still he toured, trying to make his points. He was wearing himself out. And while he was getting excellent coverage on the oil pages of the state's newspapers, he felt his message wasn't getting through to the bulk of the people.

He had another problem. Fay was hurt and angry at his long absences, his absorption in the crusade. Their quarrels upset them both.

One morning after a long night of speaking and planning, he called his office to be brought up to date on some business affairs. He told Viola Vermillion, who was still his secretary at the time, "I'm ready to give it up, Vi. I'm so Goddamned tired I feel like I'm left-handed, and I can't see that anybody's really listening except a few oilmen."

"I wouldn't be in a hurry to quit, Mr. H. I just got a mess of clippings in the mail and they're not all from the oil pages. Some are editorials supporting you."

Editorials!

He made her read every one of them. Long after he had hung up the phone a sentence from one editorial kept running through his mind: "Whether Halbouty is right or wrong or whether his campaign is a success or failure, it has been something to spark the imagination and revive the spirit of individual action. . . ." All of the editorials were laudatory and some were outright endorsements of his program.

He went back to the crusade with renewed vigor, hammering at the "archaic" regulations, laying out his blueprint for change. His religiosity, his conviction that waste was sinful as well as stupid, showed up in his speeches. He characterized Commission rules as "evils that must be obliterated." The dedicated oilman, he said, "knows he owes it to the public to make any sacrifice necessary to see that no drop of oil or cubic foot of gas—these great God-given natural resources—is ever wantonly wasted." His more cynical brethren must have smiled at that one.

He grew accustomed to the boos and catcalls, but hurled back as good as was sent. At Victoria, Texas, a burly man stood up and accused Halbouty of being a bloodsucker, ready to take the bread and meat out of a small leaseholder's mouth. "Come down off that platform, you dirty bastard, and I'll kick your ass back to Houston!" the man shouted. His companions whistled their approval.

Halbouty held up a hand for quiet and finally got it. "I was invited here to speak and I'm damned well going to do it," he said. "But when I get through, you won't have to ask me anywhere. I'll take you outside and bust your ass open like a poor boy's suitcase!"

The applause was like that on a television game show. The burly man's companions got him seated, and the noise died down. Halbouty completed his speech. He left the platform ready to fight, but the crowd surrounded him, urging him to shake hands with the burly man. The two men shook hands and hugged each other in the way men sometimes resolve such contretemps. "Oh shit," reporters heard Halbouty tell his new friend, "you probably could have taken me anyway."

Soon after Halbouty had sued the Commission, Atlantic Refining Company asked it to void an exception to Rule 37 that the Commission had granted the owner of a three-tenths-of-an-acre tract in the Normanna field in Bee County. The Commission refused and Atlantic sued, even as Halbouty had. The case reached the Supreme Court before Halbouty's. The Court ruled in Atlantic's favor, saying the one-third–two-thirds formula was an "unreasonable basis upon which to prorate gas from this reservoir. . . ."

The Court made it clear that its ruling applied to the Normanna field only, but Halbouty and his supporters were jubilant. Shortly thereafter the Halbouty case was resolved.

The Port Acres formula was invalid, the Court said, "in that it does not afford an opportunity for all parties to produce and save their fair share of the minerals or their equivalent." As far as the claim that striking down the formula would cause losses to small tract owners, the Court said, "It may be said that they drilled at their own risk." However, the Court said its decision did not foreclose on the power of the Commission to allow the holder of an exception to recover a sufficient amount of gas or oil to repay drilling costs and provide a reasonable profit.

In his public speeches Halbouty had not mentioned his possible losses under the acre-foot formula he advocated, but during oral arguments one of his lawyers, Harry Jones, Sr., of Andrews, Kurth, Campbell & Jones, told the Court: "The proof shows that Halbouty will be among those who will be hurt as a result of limiting each operator to the reserves underlying his units. Despite this, he has fought for recognition of reserves in all of the hearings involving the validity of the one-third–two-thirds formula, and is still fighting for it in this proceeding."

Justice Frank P. Culvert, who wrote the 8 to 1 decision opinion, told Jones after the hearing: "According to my calculations, Mr. Jones, our decision will cost Mr. Halbouty much more than a million dollars if his formula is substituted. That's a lot of money to pay for a principle."

The Railroad Commission's new orders for the Port Acres field were based on Halbouty's formula, and the Commission's allowables computations were based on a field map he had introduced at the original hearing. The formula, as noted earlier, accorded importance both to surface acreage and pay sand thickness. To simplify, a two-hundred-acre lease with fifty feet of pay sand was granted the same allowable as a four-hundred-acre lease with twenty-five feet of pay sand.

A small tract owner could get a special allowable giving him more gas than under the acre-foot formula if he could

show that the normal allowable would not pay out his well—
and he could show he had been refused an opportunity to pool
the tract on a fair basis.

Most small tract owners were quick to pool; the others got
their investments back and a little more.

Halbouty's loss may have exceeded a million dollars. One
oil columnist wrote:

> It is so rare that a man doggedly pursues a policy and a pro-
> gram which has the potential of costing him more than a million
> dollars, just for the purpose of standing by a principle, that the
> story deserves telling. . . .
>
> The Port Acres field was in the process of being developed.
> Halbouty had interests in some 1,500 acres. Geological informa-
> tion indicated that much of his acreage, on the east side of the
> field, was on the fringe edge where the sand thickness was at a
> minimum compared with the sand in the center of the field.
>
> So Halbouty's advisors felt obligated to warn him that if the
> pure engineering and scientific approach to the subject of spacing
> and allowables, which he was advocating, were to be adopted, he
> stood to suffer considerably. . . .
>
> By the time the issue was settled it was absolutely certain that
> Halbouty's position in the pursuit of what he calls "total conser-
> vation in the public interest" would cost him no less than $1
> million, and possibly twice that amount. . . .
>
> This may not be a rare occurrence. The oil industry is filled
> with others of similar high principles. But it does prove to a
> sometimes doubting public that there really are men who con-
> sider principle above profit, and in these days when great doubt
> is sometimes cast upon the moral attitude of modern man, such
> stories might help to revitalize our sagging faith in our fellow
> men."

And Billy G. Thompson, the Houston *Post* oil editor who
had given Halbouty the initial nudge into the fight, wrote:
"Halbouty's long fight will rank among the greatest contributions a petroleum engineer or geologist has made to the oil
industry, and to the people."

Halbouty didn't do so badly at Port Acres. When his leases finally were depleted they had yielded 2,709,177 barrels of condensate and 62,581,302 mcf of gas.

Peter Henderson, the former stockbroker, was still in the field in 1978. He had four wells producing gas, the only wells in the field, and Halbouty was receiving a royalty from them. No condensate was being produced, but outside appraisers—petroleum engineers—reported to Henderson that 27,000,000 barrels of condensate had been irretrievably lost because of too many wells and the lack of an early recycling program.

It is only now that the gas from Port Acres is selling for $2.00 per mcf. In the beginning it sold for as little as $.15 on long-term contracts. But final estimates, based on gas at the average price of $.22 per mcf and condensate at $3.50 per barrel, indicate the field produced $87,500,000 worth of hydrocarbons.

In late 1978 Henderson was preparing to drill deeper in the field to seek the Vicksburg sand, which he hoped would be laden with oil or gas. "He'd better be prepared to scratch the devil's head," Halbouty said when he heard the news.

Halbouty's victory at Port Acres started an avalanche. The Railroad Commission overhauled and improved regulations—and the Texas legislature followed suit with statutes making pooling of oilfields mandatory state-wide. And, as Halbouty had predicted, with compulsory pooling, voluntary unitization was easy to attain.

The man's energy was amazing to friend, foe, and those who simply were aware of him because of the publicity he generated. While he was reveling in the successes of the 1956-1957 drilling programs, and while he was doggedly fighting for total conservation, he found time to work for and contribute to various civic and charitable endeavors, including heading up the Houston drive for the American Cancer

Society. He wrote scientific papers. He bought banks and played the stock market.

And at the same time he set out on the wildest of wildcat ventures, one that promised him the glory he sought but brought him one of the most bitter disappointments of his life. That he failed in particular but was right in general was the only solace he could derive from the heartbreak—and heartbreak it was.

It has been recounted that at the end of World War II Halbouty flew to Canada because he believed the country was floating on an ocean of oil; that he arrived in a snowstorm and left in a snowstorm without even taking a look; that he decided to make his home base in Houston's subtropical humidity. But he never lost his interest in Canada—and he developed an abiding interest in the U.S. territory of Alaska.

He studied every book and paper he could find on Alaska's geology. There was little mention of the territory's oil and gas potential in the material he studied, but in his mind he began to envision the untamed land as the richest depository of hydrocarbons on earth.

In the mid-fifties he flew to Anchorage with Fay beside him. He hired a bush pilot, and in a relic from World War II they flew over the entire territory, with Halbouty leaning out to take pictures of everything that interested him. When he returned to Houston he was certain that Alaska was the greatest oil province yet untapped.

He said so, in speeches before fellow geologists and oilmen, and even before a four-city convocation of the Desk & Derrick Club, an organization of oil company secretaries. The secretaries were so enamored of Halbouty that the president wrote Viola Vermillion a letter she never showed Halbouty. Halbouty was marvelous, according to the letter, and all the girls were envious of Viola. She tucked away the letter, thinking that her sisters would never know his temper tantrums, the sometimes thoughtless demands.

George Hardin and his wife also made an aerial tour of Alaska, and Hardin's enthusiasm was no less than Halbouty's. Halbouty formed a new company, Halbouty Alaska Oil Company (Halasko), and set about acquiring leases in the territory, eventually acquiring drilling rights on almost two million acres.

The drill had found only one oilfield in Alaska, at Katalla, about two hundred miles southeast of Anchorage. It had been discovered in 1902 and abandoned in 1933 after producing only 154,000 barrels of oil.

The U.S. government had created a naval petroleum reserve encompassing thirty-seven thousand square miles in northern Alaska after an oil seep was found at Point Barrow in 1923. The reserve was not tested until World War II. The results had not yet been made public in the mid-fifties, but rumor had it that the reserve held from twenty to a hundred million barrels of oil.

Halbouty dreamed of finding Alaska's first great field. Most of his leases were on federal land. If he did not find oil on them before the territory achieved statehood, the leases would become state property and he would lose title. And statehood was certain.

Tommy Kelly joined the enterprise. He had graduated from Texas A&M and had gone to work as a geologist for Continental Oil Company. But when he heard that Halbouty was planning an Alaskan venture, he declared himself in. Halbouty was overjoyed. Kelly went to Anchorage to handle Halasko's day-to-day affairs.

Halbouty had spent more than three hundred thousand dollars on leases. Drilling equipment would have to be shipped from a west coast port or moved by truck over the Alcan Highway. Hunting oil in Alaska was going to cost more than twice as much as in the United States.

While Halbouty planned and worked to get started, Rich-field Oil Company, with the only drilling rig in Alaska, drilled what would be the discovery well of the Swanson River field on the Kenai Peninsula, the eastward wall of Cook Inlet.

Halbouty was in shape to move quickly to secure adjoining acreage. The discovery well had caused the first genuine flurry of excitement in Alaska, and almost every oil company had a representative in the new frontier. And almost every representative agreed that Halbouty's acreage was certain to prove a part of the new field.

The Kenai Peninsula was topped by a cruel mountain range, but between the bottom of the range and the water of Cook Inlet was a large coastal section that attracted the oil hunters. It was still rugged country, however, a heavily wooded land of rolling hills much like East Texas.

Richfield had built a road from a territorial highway to reach its Swanson River acreage. Now Hardin and Kelly, with helicopter and bulldozers, cut a road from Richfield's road to the Halbouty acreage. The area abounded in danger-ous bear and curious moose; the moose would stand all day and watch as the men cleared an area for the drillsite; the bears delighted in knocking over tents.

While Hardin and Kelly were so engaged, Halbouty had leased drilling equipment from Coastal Drilling Company of Bakersfield, California. He was so wrapped up in the project that he supervised the loading of the trucks that took the equipment to Long Beach, and he supervised the loading of the equipment for shipment by sea. The boat could not be docked at Anchorage; it went to Seward on the eastern side of the Kenai Peninsula. Halbouty was waiting for it. He took charge of transferring the equipment and supplies to trucks, and rode with the trucks to the drill site.

Drilling began only sixteen days after the equipment was loaded on the trucks in Bakersfield to be moved to Long

Beach. It was a remarkable achievement. Representatives from about thirty oil companies had converged on the drill site, and they marveled at what Halbouty and his people had done in such short time. "Hell," Halbouty said, "you major company guys would still be in a committee meeting trying to decide what to do."

It was bitter cold though it was April. The observers and off-duty workmen huddled in a shack which was never clear of tobacco smoke and smell. Most of the observers were young geologists, and sometimes it seemed as if they were more interested in Halbouty than the well. They had heard of him and read about him all of their professional lives. They believed the well would be a good one.

Halbouty, however, was in no mood to play his favorite role of professor. He was nervous, fidgety. In a sense he had no business in Alaska. Perhaps no independent did, so high was the cost of operating. He had sunk all of his ready funds into the project. He was on hand to check the final cores—he wouldn't leave that chore to others—and Hardin awaited word from him in Houston.

The bit ground down through the frozen earth, hunting the Hemlock sand, the pay sand Richfield had found. Halbouty had obtained Richfield's log of the discovery well, and he let the young geologists study it along with him. He examined each core pulled from the hole with the same intensity that had shaken him as a youth at High Island twenty-eight years earlier. The deeper the well got, the more fidgety he became.

For most of the observers the drill had not yet gone deep enough to make the cores interesting to them. And so it was that on a freezing morning Halbouty was the only scientist on the derrick floor when a fifty-foot core was pulled. As he had done in his High Island days, he tasted and smelled every inch of that fifty-foot core.

And when he tasted and smelled the last inch, he turned away from the crew and hurried behind the shack. He vomited. In the cold the vomit froze as it reached the ground. Tom Kelly found him retching, pale of face and hollow-eyed.

"What's the matter, Mike?" he asked anxiously.

"It's dry," Halbouty croaked.

"Dry?"

"That last core was the Hemlock, and it's as dry as a popcorn fart."

"But we're a long way from the Hemlock," Kelly protested. "Everybody says it's farther down the hole."

Halbouty snorted his disgust. "I was studying sands when those guys were still shitting yellow. We've got a dry hole. You can take it on down a bit more to see if there's another sand down there, but we've passed the Hemlock."

He went into the shack to pack. He said nothing to the observers, but Kelly told them Halbouty was leaving, the hole was dry. Some of the geologists couldn't believe Halbouty was serious. "You've got a thousand feet to go before the Hemlock," one said, and others echoed his opinion.

Halbouty paid them no heed. He was in despair. He left for Anchorage, leaving Kelly in charge. (The well had been staked 4½ miles from the Richfield discovery well, and subsequent drilling by Richfield would reveal that Halbouty's well was almost exactly 1,000 feet outside the limits of the field.)

Halbouty flew to Los Angeles and checked in at the Beverly Hilton hotel. He locked the door of his suite and fell across the bed.

For the first time in many years, he cried. He sobbed for the better part of an hour, ridding himself of as much of the disappointment as he could. He had not felt such pain since before the war, when he had drunk a bottle of Scotch as he sat in his car and watched while the rig came down at his third dry hole at Cedar Bayou.

Finally his sobs ceased. He phoned Hardin in Houston. "George, the son of a bitch was dry."

"I know. Tommy called me."

There was a long pause. "There's plenty of oil there, George," Halbouty said.

"Plenty," said Hardin.

"Well, let's give it another go."

"Why not?" Hardin said.

"Giving it another go" wasn't easy. Under his contract with Coastal, Halbouty had to drill ten wells with Coastal's crews and equipment, or ship the rig back to California at his expense, or pay Coastal a penalty of $375,000, the company's estimate of the cost of its returning the rig.

He squirmed and darted like a minnow in a bait bucket, trying to find a way to cut his losses. And finally he went to Union Oil Company of California to see Sam Grinsfelder, the chief geologist. Union also had leases on the Kenai Peninsula. He told Grinsfelder his problem. "The best thing for me is to move the rig back, because it won't cost me as much as the penalty," he explained. "I can move it back for a little less. But you people have all that acreage up there, Sam. You're going to have to drill it some time. Why don't you use my rig? I'll rent it to you at my cost."

"We're not ready to drill yet, Mike."

"But damn it, this rig is there! If I move it back, you're going to have to pay like hell to ship one up there. Use this one. It makes sense! It's the only rig in Alaska except Richfield's."

Grinsfelder pushed the company as much as he could. Union agreed to use the rig. It was moved to location—and Union discovered the Kenai gas field, a whopper.

Union used the rig on two more wells. By then Halbouty had lined up a number of major companies who, like him, held acreage in the Bishop Creek area. They knew little or nothing about the area and all agreed that a well to reveal the area's geology should be drilled for the sole purpose of obtaining scientific information, a common practice. The acreage was pooled and the group rented Halbouty's rig for the drilling. They found nothing to make them believe the area was productive.

Standard Oil Company of California (Socal), meantime, had obtained a half interest in Richfield's Swanson River field by putting up development money. So Halbouty talked Socal into renting the rig for several wells. In a remote basin he used the rig to drill in partnership with Paul Benedum, nephew of Mike Benedum, the greatest of the oldtime wildcatters. It was a dry hole. He drilled another remote basin and got another dry hole.

Statehood had become a fact, and Halbouty had lost most of his acreage, some of which would in later years prove highly productive to others. He had met nothing but failure in a giant area he had extolled, an area where he had marched in to lease and drill while most of the great companies held back. But even in failure, he had gained respect, it seemed. He had captured the industry's imagination, in any event; from roughneck to board chairman, oil people thrived on the romance of man against great odds.

Halbouty was determined to have one more try. He moved the Coastal rig to a lease he had obtained at West Fork, about seven miles south of the Swanson River field. The geology was good, and he proceeded to drill the deepest well ever drilled in Alaska to this point—a wildcat more than fourteen thousand feet deep.

He found gas. He had hit the gas sand going down, just below five thousand feet, but he believed that richer sands were farther down the hole. They weren't, so he brought the higher gas sand into production. But he couldn't get a gas connection; there was no pipeline to feed the gas into. He figured that some company would build a line to him if he had more production to show, and with the Coastal rig he drilled two more wells. They were dry, and proof that the discovery well had discovered very little.

Halbouty got another dry hole of sorts right in the heart of the city of Anchorage.

While Hardin was staying at a motel in Anchorage, he met its owner, a most personable promoter named Walter Hickel. Hickel told Hardin he had options on much of Anchorage's downtown area. Anchorage, he said, would grow like a weed once the oil boom got in gear.

Hardin agreed; he and Halbouty often had talked about Anchorage's future. So Hardin told Halbouty about Hickel. It was Hardin's idea that Halbouty, for a half interest, could put up the money necessary for Hickel to exercise his options. "We can sit on the property until the boom gets going good, then make a killing," Hardin said. Halbouty liked the idea. He was interested, too, that Hickel parted his hair in the middle, as Halbouty did at the time, and while Halbouty always used fountain pens containing turquoise ink, Hickel used only brown ink.

Hickel was invited to Houston for a week's stay.

It was a love feast from the beginning—two promoters promoting each other, and the world. "They kept goosing each other," Hardin told a friend, "until they thought they could build a string of pyramids in three months."

By the week's end Halbouty and Hickel decided that they weren't going to wait for a boom. They were going to erect not one but two buildings—one an eleven-story hotel, the Captain Cook, the other a fourteen-story office structure to be called the H & H Building.

They made the announcement, and their smiling photographs appeared in the Houston newspapers and on Page One of the Anchorage *Daily Times*. BIG OFFICE BUILDING IS PLANNED, shouted the *Daily Times* headline, overshadowing the crowning of Pope John XXIII, an Alaskan visit by Vice President Richard Nixon, and a couple of fairly spectacular crime capers.

Halbouty scratched up more than six hundred thousand dollars to exercise Hickel's options. Plans and specifications, drawn up by a Houston architect, cost him seventy-five thousand dollars. He had seventy-five thousand dollars worth of building steel shipped from Houston to Seattle.

All that he and Hickel needed now was fourteen million dollars, seven million dollars each, and they would become Anchorage's largest landlords.

They had promoted each other, but they couldn't promote anyone else. Once Hickel seemed close. He had found an interested investor, a Portuguese with a Hong Kong base. But Halbouty said he wouldn't accept the Portuguese as a partner.

Halbouty seemed on the verge of getting backers in New York, but Hickel called him at three o'clock in the morning. "I just had a nightmare about stampeding elephants," he told the sleepy Halbouty. "It means I ought to get out of this deal."

"You're seeing pink elephants," Halbouty said. "You're drunk!"

But Hickel wasn't drunk, and Halbouty didn't argue to preserve the partnership with his usual fire. He had lost confidence in Hickel, he explained to his staff. But there was more to it. He had lost interest in the project; it was taking too much time away from wildcatting. The dream had lasted seven months.

In later talks with Hickel, Halbouty agreed to sell his half of the property to the Alaskan on an easy-payment plan. Tommy Kelly sold the steel in Seattle for fifteen thousand dollars.

Hickel paid his debt to Halbouty. And he finally built the hotel and office building when Anchorage actually boomed.

He also became governor of Alaska and, later on, Secretary of the Interior in Richard Nixon's cabinet. He didn't stay long in Washington; he was fired for dissent, a crime in the Nixon administration.

When the partnership was dissolved, Louis Darilek, Halbouty's accountant-office manager, shook his head. "Just think," he told his office cronies, "I spent an hour running up and down Westheimer Road before I found some brown ink for that guy's fountain pen so he could sign the partnership papers!"

Like a lonely sentinel, the assembly of pipes and valves on the West Fork gas well stood through the years. For some it was a monument to a broken dream. But not to Halbouty. To him it was proof that a Lilliputian had found *something* in Gulliver's backyard.

And that proof would draw him back to Alaska time after time.

ten

BANKERS WERE SLOW to exploit the oil business. They considered it a boom-or-bust proposition and the individuals engaged in it economically unstable. A transaction, born in the great East Texas field in the 1930s' boom, changed their minds. The transaction was called "buying oil payments."

A wildcatter had a lease but no cash to finance drilling. From a friend with cash but no lease he borrowed ten thousand dollars, promising to pay the friend thirty thousand dollars out of three-quarters of the oil produced from the well. The wildcatter paid off the thirty thousand dollars within three weeks after the well was brought in.

Other wildcatters followed suit and "buying oil payments" became the pattern in East Texas oil financing. One promoter after another started getting thirty thousand dollars for ten thousand dollars in a relatively short period of time. The risk was the dry hole, the blowout or mechanical failure. There were few losses, not enough to sidetrack a trend.

Oil companies came up with a variation. A company would purchase a hundred thousand barrels of oil to be produced for the ten thousand dollars it would tender the wildcatter to help finance the drilling of his well. How good the bargain was depended on the price of oil per barrel.

Vaults of East Texas banks were bulging with deposits and paid-up mortgages, but most of their prospective borrowers wanted money to invest in oil payments. And not even the blindest banker could fail to see the great profits to be made in such short time with such small risks.

The First National Bank in Dallas was the first to put its finger in the pie by purchasing an oil payment. Others followed suit, and oil soon became as bankable a commodity as cotton, livestock, or merchandise.

(In 1931, with the East Texas boom getting underway, H. L. Hunt stepped into the Overton State Bank to borrow $5,000 for current expenses. He owned more than five thousand acres of leases in the field, he had completed one well, and had begun operating a pipeline he had built, but the bank president turned him down. Mrs. Leota Tucker, who had just banked a $30,560 lease check from Sun Oil Company, stopped Hunt and loaned him the $5,000 at 8 percent interest. Mrs. Tucker knew a good bet when she saw one. With every payment, Hunt sent her a hundred-pound sack of paper-shell pecans, a delicacy he had taken the trouble to learn Mrs. Tucker relished.)

By the 1950s, bankers had their own geologists and petroleum engineers who could estimate the value of an oilman's reserves. If the reserves were ample, an oilman had little trouble in getting a loan to continue drilling elsewhere. But most banks developed an annoying habit of wanting an oilman who borrowed say, a million dollars, to have two hundred thousand dollars on deposit with the bank. And, of course, they wanted interest.

Halbouty was painfully aware of these points, but he and

Hardin and his attorney, James Noel, had looked at banking from another angle. The oil business was great. It was their life and Halbouty and Hardin would never leave it, but it was uncertain. Banking had stability, something they needed. "We ought to get in a business where the risk is minimal so we can continue hunting oil," Halbouty said.

As often happened with Halbouty, when he thought of a project something came along to fulfill his needs. He was walking down Main Street when he bumped into W. P. (Willie) Wells. Wells was a pretty fair country-boy promoter with heavy banking experience. He was trying to put together the Continental Bank. There on the sidewalk he sold Halbouty a sizable interest. Hardin and Noel also got a piece. Halbouty added to his interest as time passed and owned 30 percent, which turned out to be the largest single chunk, and he became chairman of the board.

He didn't enjoy his role at Continental Bank. He was in constant disagreement with the other board members who could, by voting together, win all arguments. Halbouty sold his stock to Walter Mischer and Howard Terry, a couple of Houston's entrepreneurs who would place their marks on a dozen profitable enterprises.

But Halbouty had been made aware of banking's attractions. Owning a bank in rich, growing Houston was like owning a license to steal. So, even before he disposed of his Continental stock, he was receptive when Wells, abetted by Noel, told him the North Side Bank was available. Halbouty had helped Wells when Wells needed it; now Wells was pointing Halbouty to a good thing.

Halbouty was puzzled. The north side of Houston was on the decline. The city was growing to the west. Why was the North Side Bank attractive? The north side would never die, Noel told him. A strong bank would help it live and grow strong again. Wells sang the same song. They drove Halbouty around the north side, showing him the potentials for rebirth and new growth.

It was not until after Halbouty bought the controlling interest that Noel and Wells allowed him to see the inside of the
bank.

"You guys conned me," Halbouty said.

"In a good cause," said Wells.

The former owner was Damon Wells, a warehouse mag-
nate. The bank had been built like a warehouse, as if the
owner was ready to put it to a more useful purpose if the bank
failed. Halbouty was appalled. He immediately set out on a
refurbishing campaign.

More than that, he took a look at his depressed-looking
hired hands and immediately called a meeting. "I don't know
what any of you people make," he said, "but as of today,
everybody in this damned bank gets a hundred-dollar raise.
And I don't want to see any more of this damned hangdog
look around here!"

The chief teller was a man named Ernie Hand, and to
Halbouty he looked like a derelict. Hand had a large family.
Halbouty sent him to Mosk's, a Main Street emporium, and
paid the clothing bill to dress up Hand and his family. Hand
also owned a set of ill-fitting false teeth, which he never wore.
Halbouty made him go to a dentist for a better pair, and paid
the bill.

Hand never wore the new set of teeth either; he carried
them in his pocket, to be taken out only when he was informed that Halbouty was due at the bank.

It turned out later that Hand was not as poor as he looked,
and not so much hangdog as just plain weary. He had been
stealing, at about $10,000 a year, even before Halbouty took
over the bank. His guilty plea and conviction on embezzle-
ment failed to assuage Halbouty. "It's the ones you befriend
who always fuck you," he said at the time. "I know that, but
I keep on trusting people I help." Reporters, as always,
cleaned up his quotes.

In rapid order he bought control of the First National Bank

of San Angelo, built and founded the West Side National Bank in the same city, and took over the First National Bank of Paris in East Texas. And he became a founding director of the Bank of Texas, a Houston bank whose chairman was Oveta Culp Hobby, former WAAC commander, first Secretary of HEW, and publisher of the Houston *Post*.

In the meantime he had acquired the services of O. O. (Buddy) Hare, a manager in the Houston office of Arthur Andersen, one of the world's largest accounting firms. Hare, only thirty-three, was an expert at bank audits for Arthur Andersen, and he came from an old banking family in Crosby, a small town near Houston. He gained Halbouty's complete trust, and soon became not only his banking adviser but his adviser in all financial matters. With Hare at Halbouty's elbow, the banks flourished and prospered, and so did Halbouty's stock holdings.

It was well that the banks did prosper, for the time would come when he would be living not one but several steps ahead of his oil income. The strength and imagination he had poured into banking would save him from possible disaster.

James Noel left Halbouty in 1961; he was appointed to a federal judgeship by President John F. Kennedy on the recommendation of Senator Ralph Yarborough, doyen of Texas liberals. Noel had long been friends with Yarborough, and long had supported him politically. Halbouty had learned political conservatism from Miles Frank Yount, and as the years passed he moved farther to the right. In a state where Republicans were few and impotent until the first Eisenhower administration, he generally supported conservative Democrats; they usually were more friendly to the oil industry. But he and Noel had worked side by side and shared winnings and losses, never allowing their political differences to mar their relationship. This was not as simple as it appeared; Halbouty's passion extended into politics at it did into every other area of his life.

George Hardin left him in January 1961. Hardin gave as his reasons a nagging illness that had wearied him and the intense pressure he lived under as Halbouty's top associate. He had a fortune but no time to enjoy it. He wanted to fish and play golf as other men did. He could do neither while working with Halbouty.

Both of the reasons were valid. Certainly Halbouty drove himself and he drove everyone intimately associated with him. But Hardin had reasons other than those he gave. He was sensitive enough to discern that Halbouty slowly had been changing goals; he was becoming more interested in oil statesmanship than in oil finding. He was in demand as a speaker, and his speeches now were less often geological expositions than analyses of industry woes. Industry magazines printed his speeches in full and implied that he was the spokesman for the American independent oilman. Halbouty loved the role, and he was qualified for it.

As his speechmaking increased, however, there was a diminution of his promoting zeal. He wasn't as quick to round up investors to share his risks, but was inclined more often to go it alone or with fewer partners. A sales pitch to a man with a smaller income was becoming harder to make.

Halbouty also had been chiding the majors for lack of boldness in drilling untried domestic areas, and Hardin feared he was preparing to practice what he had been preaching. Hardin didn't believe Halbouty could finance such a career. And he knew that if he stayed, Halbouty's enthusiasm and magnetism would infect him to share the risks financially, and he thought such a course imprudent.

So for many reasons he departed, leaving Halbouty hurt and angry. Halbouty's possessiveness made him churlish; for months he was cool to Hardin, until his respect and affection for the man made him ashamed. With characteristic volatility, he instantly resumed the old camaraderie, and Hardin accepted it in stride.

Hardin stayed retired for a year, fishing and golfing until he wearied of fins and bogies. He returned to the oil business, holding positions with several companies until he became president of Ashland Exploration Company, the oil-finding division of Ashland Oil, Inc.

The early-day wildcatters were ready gamblers, men of wit and resourcefulness and daring. They knew little of the earth's sub-surface, but they often drilled where they liked the "lay of the land" if they couldn't find an oil seep or gas bubbles in a slough.

Halbouty was of a new breed who inherited the virtues of its elders and added to them an education in the earth sciences. And where the oldtimers were content to find oil, sell the field, and move on in search of others, men of the new breed were more likely to produce a field as their own while hunting others. They were no less wildcatters because they became independent oilmen, producing their own leases; but they did have to learn something about every phase of the industry.

There was no doubt about Halbouty's scientific brilliance, and the industry had learned that he could buy from a Dutchman and sell to a Scot and make a profit. The few times he allowed himself to be out-promoted were not in oil dealings. His mathematical bent and love of knowledge had prompted him to learn more about the industry as a whole than any but few of his fellow independents. And he could talk about it in language all could understand. He became a prominent member of such petroleum organizations as the Independent Petroleum Association of America (IPAA) and the Texas Independent Producers and Royalty Owners Association (TIPRO).

These organizations and others were in full cry for government intervention to halt the flood of imported oil after the 1956 Suez Crisis. Under proration, Texas wells could produce only eight days a month, and each well could produce no more than thirteen barrels per day. It was no better in the other oil states.

President Eisenhower finally succumbed to the pressure and instituted a system of limiting imports by what was called "voluntary controls." It failed from the moment of birth, and most independents began calling for mandatory controls. Halbouty was not among them. Voluntary, yes; mandatory, no.

Into the arena stepped a forceful figure, an elder statesman highly regarded in Texas and in government and business circles. He was Will L. Clayton, founder of Anderson Clayton & Company, world's largest cotton firm, and former Undersecretary of State. On March 17, 1958, Clayton issued a pamphlet—"What Price Oil?"—which was given wide coverage by the news media. The pamphlet said:

The high-powered drive to cut oil imports differs little from similar efforts by domestic producers of other products.

Underneath all such efforts is an understandable, human impulse to choke off competition and protect prices and profits.

Nevertheless, such attempts should be understood for what they are: Promotion of the short-term, special interest of certain producers against the National interest.

The National interest demands that such efforts be defeated.

America consumes about one-third of the energy used in all the world. Every man, woman and child is a consumer of energy in one way or another. It is highly important to every citizen and to our National economy that nothing be done to lessen our energy supply or arbitrarily raise its cost.

Proponents of restriction argue that National defense requires that we curtail our dependence on foreign oil.

How long do they expect World War III to last?

The President's Materiels Policy Commission, headed by Mr.

William Paley, states that by 1975 consumption of petroleum products may be expected to double in the United States, triple in Europe and quadruple elsewhere.

The Paley Commission further states: "The energy economy of the United States has prospered on the basis of using the cheapest available fuels and can prosper most in the future if our import policy continues to permit oil consumers to have access to the lowest cost sources consistent with security. Geological and economic conditions throughout the world favor an increasing reliance on imports to meet a considerable part of the future growth of United States consumption, even though the United States production of oil can also be expected to grow. Consumption is expected to increase more rapidly than production, so as to leave room both for increasing imports and a healthy domestic petroleum industry."

We should never forget that the United States has only about 20 percent of the proven oil reserves of the world, whereas we are consuming over half of the present production of oil in the world.

How long will we go on protecting our minorities at the expense of all our citizens? How long will we continue an archaic policy which angers our allies and the Free World generally, causing bitterness and retaliation?

The Russians are smart. They roam around the world, offering trade. We give away some millions here and some there. The Russians give little, but they trade. No self-respecting people want charity; they want to earn their way.

Is it any wonder that we are losing the Cold War?

These are tragic times for all the peoples of the world. As long as the United States maintains its ability to destroy Russia in event Russia starts destroying the United States, World War III is unlikely. But we could, nevertheless, lose almost everything that is dear to us, except life itself, without a shot being fired.

To seize the initiative in the Cold War and turn defeat into victory, we must first make ourselves worthy of the leadership of the Free World. We will never do that so long as we continue to act in the short-term, special interest of our minority groups against the National interest and against the needs and interests of our allies and other nations of the Free World.

Our oil imports come partly from Venezuela (buyer annually of $1 billion of American goods, the economic equivalent of 250,000 American jobs), partly from Canada (our best customer in the world), partly from the Middle East (just now the powder keg of the world).

Are we going to make all these areas mad just to maintain high prices and big profits for oil producers? If so, we are headed down a road which leads to disunity in the Free World and its eventual defeat.

Perhaps because his company was an international trader in more products than cotton, Clayton had taken the viewpoint of the international oil cartel. But he obviously knew little of the history of oil importation, or ignored it, and his most positive assertions left him open for counterattacks. But such was his stature that Congressional proponents of imports limitation were silent. Most of the independents were silent because they knew no more about the intricacies of the question than Clayton appeared to know.

Halbouty, at this stage, did not know all he wanted to know about the problem, but he felt he knew enough to more than adequately reply to Clayton. And he was as angry as he had ever been. He held off reporters until he could quickly scratch out a statement. It made Page One of many newspapers and was printed in full in *Oil* Magazine.

First, let me say that I speak for no company, no association or group. I speak only for myself. As an independent producer of oil and gas I am in a position to do so.

I gathered that Mr. Clayton thinks that this country should permit an unlimited supply of foreign oil to come into the United States without regard to the domestic producing industry.

He said the National interest demands that the efforts to cut oil imports should be defeated. The opposite is true. Oil imports are now supplanting domestic production. They should be reduced to the point where they supplement. When they begin to supplant, they begin to destroy the initiative necessary to a healthy domestic industry. That, in turn, kills off required risk capital, puts men out of work and causes economic distress, if not, in fact, upheaval.

Mr. Clayton's own words defeat his intended purpose, in my

opinion. He says (in defending unlimited oil imports) that "Every man, woman and child is a consumer of energy in one way or another," and that "it is highly important to every citizen and to our national economy that nothing be done to lessen our energy supply or arbitrarily raise its cost."

I most wholeheartedly agree with that. But this nation must not become dependent on outside sources for its energy fuels, as it would surely cause it to lose its position as a world power and cause its people to suffer as a consequence. If imports are limited and not supplanting domestic production, we can maintain a healthy domestic oil industry always capable of meeting the demands of our people without ever having to depend on any monarch, dictator, commissar or foreign cartel. . . .

With unlimited foreign imports, the market for domestic production will be so curtailed that no new reserves will be found, and we will become a "have not" nation. The hunt for new reserves must go on constantly since it takes time to find and develop such reserves. In times of need such as in World War II, the nation must have available the immediate productive capacity with which to defend itself. Only a healthy domestic industry which annually discovers as much oil as it produces can assure the nation of this potential. . . .

I contend that the best service this nation can do for itself and its allies in the Free World is to stay strong economically. If we impair the economic health of the industry that provides us with two-thirds of our energy fuel, I believe it will begin to undermine the entire economic structure. That could be what is happening right now.

Halbouty wasn't completely correct in his reply, but he was right more often than wrong, and considerably righter than Will Clayton. In any event, he established himself as the man to call on when the enemy stood hammering at the gates. As for Clayton, he issued no more pamphlets.

In the late 1920s the three great marketers of foreign oil formed a cartel, hoping to wax fat on the world's growing demand for oil. They were Standard Oil Company of New

Jersey, Shell and Anglo-Persian Oil Company. In later years Standard would become Exxon (Humble Oil & Refining Company, a Standard subsidiary mentioned earlier, would become Exxon USA). Shell would maintain its name into the present. Anglo-Persian would become Anglo-Iranian and finally British Petroleum.)

The three giants decided to sell cheap oil to Europe and other areas outside the United States at United States prices. They were producing the cheap oil in Venezuela, the Middle East, and what is now Indonesia. It was to their advantage, then, to keep the price of United States oil as high as possible.

In addition to charging United States prices for the cheap oil, the companies decided to add a fake shipping charge as if the oil had come from the Gulf Coast. Thus Europeans, for example, would be paying dearly for oil from the nearby Middle East that was produced for a few cents a barrel, and could be delivered for only a fraction of the Gulf Coast shipping cost.

For the scheme to work, production in the United States had to be controlled. The Depression and great discoveries in the United States, particularly the East Texas field, saw to that. There was a great glut of oil. Conservation programs were instituted in the United States to prevent waste—and they were based on market demand. Fields in Texas and other oil states could only produce the amount of oil the majors ordered each month. With proration, the price of domestic oil rose, even during the Depression.

So the scheme worked. The cartel grew until it finally settled down to what became known as the Seven Sisters—the companies that are now Exxon, Shell, British Petroleum, Standard Oil of California, Gulf, Texaco, and Mobil. Many of the other plans of the cartel died, but the Gulf-plus, as it was called, persisted. European countries and others outside the United States continued to pay the United States price for cheap oil plus a fake shipping charge.

And the cartel continued to flood the United States with cheap oil, charging United States prices for it when it was transformed into gasoline and other products. Gasoline at the pump in Wisconsin cost the same whether it was made from Gulf Coast oil at $3.00 a barrel or Middle East oil at $1.65 a barrel. And more and more it was made from Middle East oil.

The independents' oil, then, was competing against imported oil plus the domestic oil the majors produced from their own acreage. The independent was getting top dollar for his oil, thanks to the cartel and proration, but he wasn't allowed to sell much of it. With a small market, there was little incentive to drill more wells, hunt new fields. And the majors had little reason to drill more domestic wells or hunt new domestic fields, either, with foreign oil so plentiful and easy to obtain. They cut their staffs of geologists and other oil-finders to the bone.

Perhaps the cartel, and proration, permitted American consumers to obtain abundant supplies of oil at low prices, as Clayton and others claimed.

Fifteen years from the time of Clayton's pamphlet and Halbouty's reply, the American people would be in a position to wonder if the price was right.

During those fifteen years, Halbouty would spend most of his time and much of his fortune shouting a message no one wanted to hear.

eleven

HALBOUTY had no real desire to attack the majors. He had maintained friendly relations with most of them over the years and, at times, had defended their positions on domestic issues. But more, he staunchly believed that the entire industry should present a solid front to the American Congress; he feared the threat of federal control like the devil fears holy water. He was hesitant to crack the solidarity.

But the vision he held of the future troubled him. The American people had to be made to peer into the future with him. If they would, perhaps catastrophe could be avoided. To explain the future he had to explain the present; if chips fell on the majors' doorstep, so be it.

Halbouty made his first important speech on the subject in Los Angeles on November 3, 1960, before the thirty-seventh annual meeting of the American Association of Petroleum Geologists. The huge convention room in the Ambassador Hotel was packed with delegates. The news media was present in full force; Texas reporters had tipped-off their West Coast buddies that Halbouty was "hot copy."

By now voluntary import controls had become mandatory controls The majors had cut the price they paid for oil. The Russians were getting active in the oil market. Industry efforts to force Congress to deregulate natural gas prices were growing stronger. Halbouty discussed all these. And he praised the American majors for their courage in going overseas to drill and staying to produce oil under the threat of Communism.

But he made one statement that he would make over and over in the following years, varying the verbiage but never the content:

> History has demonstrated through such crises as World War II and Suez that foreign governments by deliberate action can deny the United States the use of foreign oil. It is obvious that if we become dependent upon foreign sources, our security and peacetime welfare are at the mercy of action completely beyond our control.

He came back to the theme again, this time tying it to a startling suggestion that the majors use some of the profits from cheap foreign oil, $1.00 a barrel, to explore for the domestic oil and gas that surely would be needed in the future.

Halbouty also offered a suggestion made by the American Association of Oilwell Drilling Contractors. The organization had just reported that in the first eight months of 1960, imports were sixty thousand barrels a day above the average for the same period of 1959, and domestic production was down seventy-seven thousand barrels.

> The association suggested that imports be reduced and limited in 1961 to enable domestic producers to regain their share of the United States market lost since the end of the Suez crisis, and that thereafter domestic producers be permitted to furnish the major portion of the increase in United States demand. I think it is a most reasonable suggestion. . . .

Independents drill approximately 75 percent of all wells drilled in the United States and almost 85 percent of the exploratory or wildcat wells. The only incentive that keeps many independents working to drill one wildcat after another is the hope of finding "the big one." In that regard he is dependent on venture capital from sources outside the industry. This capital is dwindling because of the present status of our industry, so that fewer wells are being drilled and less new oil is being found. . . .

We are in trouble when we have a surplus of domestic oil and then add 1.75 million barrels of oil imports a day. That amount of imports is in excess of our present reserve production capacity of 1.5 million barrels daily to meet our national defense requirements. In other words, if there were no imports at all, we would be about where we should be relative to crude oil supply in this country.

As he ended his prepared speech, he picked up some additional notes he had written. Now he spoke with a special earnestness, and there was a touch of anger in his voice.

I want to end this presentation by making it perfectly clear that, in my opinion, this country is reaching toward a severe economic crisis, as well as an imposition on our national security, by our not doing everything possible to increase our domestic production now. To continue along the downward exploratory curve which we are now experiencing will surely result in economic chaos ten years from now. The impact of an energy shortage in this country would be absolutely disastrous.

Unless there is an appreciable and sustained turn-around in our exploratory activities, I can safely predict that between now and 1975 we will have an energy crisis in this country which will cause repercussions throughout the width and breadth of this great nation of ours like a devastating earthquake.

It is appalling to me that the American people can be so apathetic to what is so obvious to some of us in the industry. The people of this country just don't care. They are not experiencing shortages now, and evidently they care less about what will happen in the future.

Some of these days the shortage will catch up with us, and then the people will say, "The industry is to blame. Why weren't we told?"

He pounded a fist against an open palm. *"Well, I'm telling 'em now!"*

He had hoped for widespread news coverage, and he got it. The wire services picked up the story and sent it out across the land. But the story, in almost every instance, wound up on the business pages of newspapers, and without the warning he had sounded for the American people. His angry indictment of public apathy, however, found its way into much of the trade press, but with little or no editorial comment.

He was disappointed. "This whole fucking business is full of heartaches," he told Louis Darilek, his office manager. "I just drilled a dry hole in what I hoped was productive territory. I guess I'll have to keep trying."

Said Darilek: "They'll pay attention when they drive in some place and say 'Fill 'er up,' and some pump jockey says, 'With what?'"

Halbouty's reply to Clayton started a fifteen-year Count Down. Now Darilek's statement was thirteen years before the event.

While the majors publicly acted as if they were unaware of Halbouty's Los Angeles speech, some company officials privately scolded him. To their surprise, he didn't argue with them; he was saving his fire. Some of his associates warned him that he was fouling his own nest; he shrugged them off.

He had taken fire from some independents when he had fought against mandatory controls. He had not wanted mandatory controls because he feared any federal government intervention in the industry. Further, he knew that the State Department had been using oil as a political tool for years in negotiating with the Arab states and others. Mandatory controls would give the department a stronger hand to play, and the independents would suffer.

And the flow of imports did increase under mandatory controls. There was a movement among independents, then, for what was called "legislative controls"—enactment by Congress of laws setting import restrictions.

Halbouty almost went up in smoke. Mandatory controls, as distasteful as they were, could be amended, revised, or even rescinded, he told his brethren. "Only an act of God can change an act of Congress, as far as the oil business is concerned," he counseled. He was not gentle. His scorn made him enemies in his own camp, but his influence and rhetoric caused the movement to quietly die.

Halbouty spoke wherever he could find an audience, as he had done in his struggle with the Texas Railroad Commission. This time, however, he crisscrossed the nation. Months would turn into years. He would grow weary and bitter at times. But his natural optimism would erase them both, and he would carry on.

He urged each industry organization he addressed to join him in his missionary effort. He softened his remarks about the majors and appealed to them to unite with the independents in working out a solution to what he considered "one of the paramount problems of the era." He called on government to include independents in its import thinking and decision-making.

To be sure, he spoke of other matters in his almost endless rounds. He urged geologists and geophysicists to lay aside their professional jealousies and work as a team in the search for hydrocarbons, for example. Time after time he took the industry apart as if it were a fine automobile and held up the moving parts for his audience to examine. And there was a warm love of the industry running through his words, an exasperated affection, which kept him from being a common scold. Darrel Royal once described himself as the "pride coach" when he was football mentor at the University of Texas. Halbouty was the industry's "pride coach."

But he could not long ignore the vision that haunted him. He told one group:

Many contend that Middle East oil costs only pennies a barrel to find and produce. This is an illusion. Their figuring does not include the international accommodations which are costly to our taxpayers and have to be made at every turn. It does not include the cost of tremendous damage to our domestic industry, which has seen exploration virtually limited to offshore areas, the disappearance of thousands of independent domestic operators, and the exodus of experienced and needed drilling contractors from the energy picture. It does not include a thousand other untold costs to our economy and our security, including our losses in the balance of payments.

I contend that oil from the Middle East is the most expensive commodity we have in the world today!

As usual, when his remarks were printed they appeared on the oil pages of newspapers. "Oil editors outnumber steel, flour and pharmaceutical editors on daily newspapers, but few people outside the industry ever read the oil pages," he complained in one speech. He said he knew why. "The general reader learned long ago that there was little on the oil pages to interest him; the industry loads down oil editors with company handouts in which no one is interested except the company that made the release, and, ironically, only a few people in the company are interested."

He made a decided effort to break out of the oil pages by speaking to service clubs and organizations, but almost invariably it was a business writer who was sent to cover the meeting. And he learned also that service clubs and organizations were not as enthusiastic about Halbouty the "doomsayer" as they had been about Halbouty the rejoicer in the free enterprise system.

The occasions when the applause was warmest were when he pointed to Russia. Apparently his audiences could indentify with anything anti-Soviet. "Russia," he said in 1964,

is now trying to prove to the Europeans that all of Europe must eventually depend on the Soviets for its energy supply. Even over political conference tables, the Russians make it a point to inform the heads of European countries that their future oil supply is in the hands of the Soviets.

In the past, Europe has always looked to the United States for oil when its supply from other oil-producing countries was cut off, as in the Suez crisis. Yet, Europe now is saying that it wants a diversification of "sources of supply" to insure itself of oil when a crisis arises. This is a smooth way of telling us that Russia is now more than a hole card to them; that now the European countries will trade with Russia, and their dependence on the United States is over. . . .

They forget how suddenly the supply from these "other sources" can be choked off. In today's volatile world it is entirely possible for a catastrophe to occur elsewhere which would in turn bring disaster to Europe.

Just think what would happen to the European countries if Middle East oil is again blocked at Suez, if the North African supply is stopped by insurrection, if South American oil is held up by Castro and his Communist allies, or if Russia decides to slash Europe's oil supply to teach it a lesson.

Then he would turn to the message he wanted to deliver: that there was more oil and gas in the United States than had been produced but it wasn't being discovered because of our growing dependence on foreign oil; that the Middle East was not a dependable source of supply because of the area's political volatility and the possibility of a Russian takeover; that we should begin now, today, to find and develop oil and gas fields in the United States to make the country independent of the actions of others.

He was fierce in his defense of independent oilmen, old breed and new, and particularly when it was suggested that the independent no longer had a place in the industry. "A few years ago," he said, "there were forty-two thousand independent explorers and producers in this country. Today that number has shrunk to ten thousand! The day is not far off when not even that number will be able to stay in business. But those who do will be as important, if not more so, than any other segment of the business because they will continue to find a large proportion of the future oil."

Many of the explorers who had fallen by the wayside, he said,

> were maligned and castigated by the press and the politicians. But each of them made a greater contribution to his country than all of their critics combined.
>
> Certainly they were show-offs. They drank whiskey and spent money and carried on like an army of Coal-Oil Johnnies. But they were a different breed of men. Most of them were laboring men who had more than average intelligence and a will to work as hard as they played and were blithely unaware of the odds they faced.
>
> They never disturbed our economy because most of them lost their fortunes as fast as they made them. But in doing this, they were finding the fuel to provide the energy for a new way of life. There is not another breed of men on earth who could have or would have done what they did. My hat has always been off to them, and my head is bowed at their passing. . . .
>
> No school can produce an independent oilman. The universities turn out geologists, engineers, lawyers, accountants, and others; some of these become independent oilmen. But no college or university has a course in intestinal fortitude or hard work or how to recover again and again from one heartbreak after another.
>
> And corporations seldom develop such men even with all their screening, selecting, supervising, and training. The thinking and the make-up of independents place them in a distinctive class.
>
> The American independent oilman is *not* going to die out!

The time was coming, he said, when America would need the wildcatter more than ever. "That's the day, probably before 1975, when the United States will have to start playing catch-up, the day someone, some group, some country, shuts down the valve on foreign oil."

He was tired of statements that the United States had little hope of sustaining itself on domestic oil.

Hundreds of millions of acres of sedimentary land in this country have never been explored at all! The proven and productive sedimentary basins still have to be explored deeper—even below twenty-five thousand feet.

No man can say with accuracy what the earth holds in the way of hydrocarbons. Too many have tried it, many of them as distinguished and as learned in their day as present prognosticators. They have all goofed!

The learned men were saying we were out of oil when Pattillo Higgins and Anthony Lucas found Spindletop. The experts were stating unconditionally that we would have to depend on oil from Venezuela when old Dad Joiner found the great East Texas field.

Who knows where or when or how many fields we might find with great reserves like the King Ranch, Katy, Carthage or Hastings; or Signal Hill, Kettleman Hills, or even the Panhandle or Hugoton?

The geologic conditions in this country do not rule out such possibilities. All that is required is sound scientific exploration.

It seemed that in every passing year the Russians did something to harden his conviction that their wooing of the Arab states was a courtship designed to deprive the West of Middle East oil. He talked about it constantly.

And in June 1967 Arab and Israeli guns flashed in the Six Day War. The flow of oil was shut off. As it had done in the Suez Crisis, the United States saved Europe from an oil famine, supplying ninety million barrels over and above normal production while the new crisis lasted . . . an amount, interestingly enough, exactly equal to the amount the United States provided to win World War I.

Thoughout the crisis the Soviets urged the Arabs to use oil to punish the United States and other western powers friendly to the Israelis. The Arabs resisted, for reasons of their own, and the flow of Middle East oil resumed.

The resumption of the Arab oil flow did not allay Halbouty's fears. He was surprised at it because he had fully expected the Arabs to succumb to Soviet blandishments. He had paced his office floor during the brief conflict, his ears tuned to the radio, his eyes on his television set. He had called dozens of friends, seeking inside information but rarely finding it. When the crisis had passed he called persons he considered influential and pleaded that they not be misled by the Arab resistance to the Soviet plan. He discussed the plan in his next speech.

Said Halbouty:

> The Soviet's program appears to be one of controlling 90 percent of the world's liquid hydrocarbon reserves, dumping them on the world market at whatever price it takes to make others dependent on them, and then giving their customers the choice to communize or return to the economy and defense vulnerability of the nineteenth century.
>
> This plan of world control can be stopped, but not as long as western nations ignore the pointers, or as long as our government and our citizens do all in their power to destroy the most vital and important industry in the history of world progress.

If the government and the people wanted to ignore Russia's intentions, they were still in trouble. "The Middle East oil states can harm us without Soviet help," he said. "Every day they move closer to nationalizing their oil and, in my opinion, that brings us ever closer to a time when they can shut us off, for any reason, and hang us out to dry. . . ."

Venezuela had pointed the way for the other foreign oil-producing countries to improve their positions with the multinational oil companies. Since the 1920s, the Venezuelans had received only 7 percent of the oil produced. A short-lived democratic government, at the end of World War II, passed a 50-50 law which gave the government, through taxes and other devices, a 50 percent share. A succession of administrations found no reason to disturb the 50-50 law.

The other foreign oil-producing countries, as time went by, incorporated the essentials of the 50-50 law into their dealings with the multinationals and, as their economic thinking matured, demanded and sometimes received even more concessions.

The multinationals received another blow after World War II, this time from the United States government and European countries the United States was aiding under the Marshall Plan. Under their Gulf-plus program, the companies were charging American taxpayers the Gulf Coast price of oil plus Gulf shipping charges for cheaper nearby Middle East oil they were sending into the rebuilding nations. Under pressure from angry Marshall Plan administrators and the State Department, the companies agreed to drop the phony shipping charge for the duration of the economic recovery program.

Despite their concessions to foreign oil-producing countries, American importers still found foreign oil more profitable than domestic oil, so the flow from the Middle East and elsewhere never slackened except in times of crisis.

And the majors knew just what to do when a foreign producer got too demanding. In 1951, after a dispute about the application of the 50-50 profit rule, Iran nationalized its oil industry. The companies immediately clamped a total worldwide boycott on Iranian oil, an embargo that lasted for more than three years. It was a cruel blow to the Iranian economy, one from which the country was slow to recover.

It also was a harsh warning to the other foreign producers. But the foreign producers kept on taking tiny steps toward cooperation against the companies, despite their fears.

And in 1960, the year of Halbouty's Los Angeles speech, the Organization of Petroleum Exporting Countries (OPEC) was formed. The acronym in time would become as familiar to Americans as JFK, RFK, or even M*A*S*H.

Halbouty spoke with such certitude that perhaps only James Clark realized that doubts ofttimes pinched his tongue.

Basically, he could not help but praise the majors for drilling and producing in foreign lands; it was as much a part of the free enterprise system as opening and operating a filling station in Cut and Shoot, Texas. And he had become convinced that the American companies had gone abroad at the behest of the U.S. government to prevent European companies from controlling foreign sources of oil supply. He felt bringing foreign oil into the United States was important and helpful—economically at home, politically abroad.

But he knew that potential fields in the United States were being ignored by the majors while the independents could not afford to drill them because there was no domestic market to speak of. It might be well for the American consumer to obtain all the foreign oil he needed while conserving his own, provided the world remained stable. But history said the world would not remain stable. History said the flow of foreign oil could be shut off without notice.

"If and when that happens," he said in a note to Clark in 1962, "we'll be at their mercy, one way or another. If they keep the valves closed to us, we'll literally starve for oil because we haven't maintained our domestic reserves. If they open the valves after a time, it may be at such a cost to us to make us a second-class power. Just think how long it will take us to find and produce the oil we have in this country, starting from scratch!

"But hell, Jim, maybe I don't know what I'm talking about. I've drilled my share of dry holes, haven't I? Maybe I've misjudged the situation from the word go. You can tell me at lunch Wednesday if you think I'm acting like a mule's ass. . . ."

As "pride coach" for his professions, Halbouty was a Distinguished Lecturer for the Society of Petroleum Engineers in 1964 and 1965, and in 1965 and 1966 was a Distinguished

Lecturer for the American Association of Petroleum Geologists, the world's largest association of earth scientists. In 1965 he received the Texas Mid-Continent Oil & Gas Association's Distinguished Service Award.

He neglected his business in preparing and delivering his speeches, traveling from one state to another, one country to another. For every invitation to speak he accepted, he had to turn down a score.

He was a dramatic orator, a handsome, commanding figure of a man. Whether he was telling his colleagues how to improve their work and profession, denouncing Congress or some Congressman, or warning of the oil shortage he foresaw, he gave the impression of having a profound knowledge of and sincere fascination with his subject. He had never lost his ability to study; he was a compulsive reader, and when he read of something that interested him he liked to go to the scene and examine it.

This period of his life was crowned when the AAPG elected him as its president for the 1966–1967 term. ("If 'Prexy' Walton hadn't loaned me fifty dollars, and if I hadn't worked the shirt off my back to pay him back, I'd still be selling bananas in Beaumont.")

The AAPG had members in seventy-five countries at the time, and many of them outside the United States felt neglected. The Canadians, in particular, had begun talking secession from the AAPG.

With Humble geologist Merrill Haas, another AAPG official, Halbouty flew to Calgary to meet with the dissidents. The Canadians were cool. "If they'd had horsewhips, they would have driven us back across the border," Halbouty later told his staff. He was apologetic to his hosts. He knew he was the first AAPG president ever to visit the Canadians, he said, and he knew they had been ignored. He was in Calgary to start making amends.

The Canadians were not to be appeased. "Listen," said one

geologist, "you people promised to hold an annual convention up here, and you never did. You made that promise ten years ago, at least. You never do anything for us."

Halbouty pointed a finger at him. "I've been trying to softsoap you, and I knew it was a mistake before I started it. Now I want to tell you, if you want to secede, you secede. And it will be your loss, not ours. But I'm the president now, and I'm telling you that you're going to get a convention. Now when the hell do you want it?"

"In 1970."

"You got it," Halbouty said. "Now let's talk about our real problems."

"You can't guarantee it," said the Canadian.

Halbouty grinned at him. "I'll railroad it through."

Someone laughed, constructive talk began, and secession talk faded.

The Canadian experience prompted Halbouty to make other journeys to shore up AAPG outposts . . . to Tripoli, the Hague, Mexico City, Caracas, and London. While in London he spoke before the prestigious Institute of Petroleum. After the meeting he dined with Dan Ion of British Petroleum, the Institute chairman. A few wells were being drilled in the southern portion of the North Sea at the time. Halbouty and Ion agreed that the North Sea held vast potential as an oil and gas reservoir but that the current drilling was too far south.

From their talk grew a convocation of European and African geologists in Brighton, England, where the earth scientists shared their knowledge and theories. It is commonly agreed that their discussions hastened the movement of drilling rigs northward in the North Sea, to where great deposits of hydrocarbons would be found.

John Moody, a senior vice president of Mobil Oil, later a geological consultant, served on the AAPG executive committee during Halbouty's tenure as president. "It was one of the—if not *the*—most productive administrations," Moody said at an AAPG annual meeting. "Mike's contributions to and achievements for our association were outstanding. Decisions and implementations he made as president have had, are still having, and will have in the future, far-reaching effects, not only on AAPG as an association but on our profession as well."

He had mellowed but little. He still was as vain as a pimp with a stable of whores. He still was brusque, suffering fools not at all and others when it pleased him. He had a battery of the most modern electronic gadgets on his desk to summon aid, but he still shouted for his secretary or anyone else he needed.

There was, however, one noticeable change in his emotional characteristics; he could laugh more easily at himself, as if he had become aware of his foibles. He could shake his head and say, "Jesus, I was an arrogant bastard back in those days." The statement would be followed by a knowing, sardonic chuckle, as if to say, "I'm *still* pretty damned arrogant."

He had been asked to chair an important committee for the National Academy of Sciences, and he was working in his office one Saturday morning, trying to locate a member who was to serve with him. The man was associated with a western university. Halbouty asked his secretary, Mary Stewart, to get the man on the phone.

The university switchboard was closed. Stewart got a residence number from information and dialed. Halbouty took the phone from her. "I'll talk to him," he said. "You get on your phone and take notes."

"George!" Halbouty said cheerily. "This is Mike Halbouty in Houston! How you doing?"

"Who?"

"Mike Halbouty! I want you to serve on a special subcommittee, George."

"Who did you say this was?"

"Mike Halbouty, damn it! Don't you remember us talking about the committee a couple of weeks ago? On water technologies?"

"Mike who?"

"Halbouty! H-A-L-B-O-U-T-Y. You're not drinking, are you? It's pretty damned early in the day for that, George."

"I don't believe I know you."

Pause. "Say, you *are* a geologist, aren't you?"

"No, I drive a beer truck."

Halbouty grunted. "Well, I'll be a son of a bitch." He chuckled. "Tell you what, George. How would you like to be on the committee, anyway?"

"I don't think so."

"All right, George," Halbouty said, still chuckling. "Maybe next time."

twelve

Meanwhile, he was still in the oil business. In the fields he had discovered, he was drilling about seventy-five wells per year—about three every two weeks—making him one of the busiest operators on the Gulf Coast, including major companies. These were development wells, and there were few dry holes among them.

On the wildcat front, however, he was drilling more dry holes than producers, and some of the wildcat failures were spectacular. And costly.

Halbouty was hunting "giants." He was prepared to drill deeper and farther away from established production. And often, though not always, he was forced to spend more of his own money on leasing and drilling than he had in the past because partners were harder to find for such undertakings.

Possibly the most spectacular failure of this period was on the Pescadito dome, a massive uplift in Webb County about fifteen miles from the Mexican border. The largest salt domes in the Gulf Coast area were from four to ten miles across. Pescadito dome was fifteen miles long and nine miles wide, 135 square miles. It lay within the borders of a great ranch owned by O. W. Killam, an old wildcatter who had found oil aplenty in South Texas.

Two Houston wildcatters, Sid Brewster and Bill Bartle, had drilled a dry hole below 8,000 feet on the crest of the dome. They were followed by three other Houston wildcatters, Noble and Wilbur Ginther and Howard Warren, who had drilled a dry hole to 15,107 feet. It had cost them about six hundred thousand dollars.

Halbouty couldn't stay away from Pescadito; the industry didn't expect him to. Aware of his interest in the dome, Ginther, Warren & Ginther approached him. They couldn't get any help to drill another well, they said. If Halbouty would come in with them instead of drilling for himself, his reputation would bring Gulf into the fold. And Killam himself would take a piece of the action.

Halbouty was not immune to flattery. He and George Clements, GWG's geologist, agreed that a well should be drilled to the northwest of the GWG dry hole, and to a greater depth. The bit would be seeking the Edwards Lime, a chalky formation that had proved productive in scattered wells some twenty miles to the north.

Halbouty agreed to participate in the venture, and Gulf joined the action. Halbouty was optimistic, as always. Because the operators expected to find extreme temperatures down-hole, special steel and rubber equipment was used to combat heat. Special, heavier drilling mud was employed.

The well blew out at 14,004 feet and caught fire. Fortunately, the wind was blowing from the right direction and it blew the gas away. The fire was confined to the mud pit and burned for only twenty minutes. The gas generating the blowout was only a small pocket.

The pressure was controlled with heavier mud, and drilling

resumed. The drill pipe stuck. Drillers had to sidetrack the stuck pipe and drill around it. Then the bit struck a peculiar shale that would slough off in chunks and stop the bit from turning in the hole. The drill pipe broke. The broken pipe was fished from the hole and drilling resumed.

The drilling mud was so heavy that the shale floated in it, and chunks of shale even washed up into the mud pit. The temperature at the bottom of the hole reached 419 degrees. Mud from the bottom was so hot that it burned out rubber return hoses. Steel hoses were substituted.

The last 3,000 feet were drilled with a diamond bit, an expensive accessory. At 19,503 feet they had found nothing they could identify because it was too hot to take a core. They cussed and discussed and finally said to hell with it.

Halbouty participated with GWG in drilling seven more Pescadito dome wells. Three of them were completed as gas producers from shallower depths. By 1978 they had paid back about half of the money he poured into the dome. One well was still producing. Others who moved onto the dome later spent their money and left. The bit never found the Edwards Lime.

The Katy field, to the west of Houston, was a giant, producing gas and oil from shallow Yegua sands. It was surrounded by small salt domes, but its chief geological feature was a large anticlinal structure. In 1937, Pan American had drilled a deep test on the structure in a search for the Wilcox sand. The bit had found the Wilcox below eleven thousand feet, but Pan American geologists concluded the sand was too "tight" to produce. The well was abandoned. Meanwhile, the Yegua sands gave up their riches easily.

Halbouty had an idea about the Wilcox sand. His studies convinced him that if he drilled down-structure to the west, the Wilcox would be looser—and productive—if he found it.

With Hardin gone, Halbouty's right-hand man now was Tom Barber, the exploration chief who had so well represented Pan American at Port Acres. Barber shared Halbouty's beliefs and optimism about Katy.

Halbouty built up a lease block of four thousand acres, but the more he studied the area the more acreage he wanted. He continued leasing, and wound up with seventeen thousand acres. "The only lease Halbouty wants," said a weary landman, "is the lease right next to the one he has."

Halbouty acquired several partners in the venture, including Houston Natural Gas Corporation, and drilling began. The bit found the Wilcox, but it was bearing salt water. So were other wells that followed on his acreage.

But Halbouty's wells prompted Pan American to take another look at the 1937 well where the Wilcox had been deemed too tight to produce. This time it wasn't too tight and Pan American obtained production. Halbouty's acreage was out of the play entirely.

While Halbouty was still reeling from this setback, he drilled a dry hole in West Texas. Shortly thereafter another wildcatter discovered a gas field nearby. Since Halbouty had given up his leases, the wildcatter moved onto the acreage and dropped his bit into Halbouty's dry hole. He drilled five hundred feet deeper—and found gas.

Still hunting "giants," Halbouty moved into St. Landry Parish, Louisiana, where the Opelousas salt dome had been productive for years. He was not interested in the dome itself, but in a satellite feature, a huge fault across the dome's south flank. His geology and a geophysical survey prompted him to hustle up leases until his block was eleven miles long and five miles wide!

Halbouty and Barber believed they would find a major reservoir in the Middle Wilcox sand. What they found was production from the Upper Wilcox; the Middle Wilcox was too shaley to hold oil. The rig was moved to the west. This time they found no production in the Upper Wilcox, but found oil in the shallower Sparta sand. The rig was moved again. This time the bit found oil in the even shallower Frio sand, and nowhere else.

Halbouty had to content himself with drilling and producing the shallower sands on his acreage. They were still producing in 1978, but the field was not the "giant" he had envisioned.

He went back to South Texas, this time to Zapata County, where a Union Producing Company wildcat had opened the Escobas field. Halbouty quickly acquired 640 acres offsetting the Union wildcat, and to the south, where he located another structure, he grabbed another lease block.

Using the offsetting acreage as solid bait and the acreage to the south as the "romance," he made a partnership deal with Union Carbide Company to drill both blocks.

A well drilled on the offsetting acreage, the solid bait, was dry. A well on the southern acreage opened the South Escobas field. Three other productive wells were drilled on the block, but they were short-lived.

In Victoria County Halbouty found the buried channel of a bygone underground stream that had cut a swath through the Upper Wilcox. It came to him that the eroded sand had been pushed along and dumped somewhere like a heaping spoonful of sugar. He located the area where the spoonful was likely to be, if it existed. This was wildcatting of the rankest sort; there was no structure of any kind to indicate an oil entrapment, only the evidence of the long ago stream and the spoonful-of-sugar deduction.

He made his case well enough to entice four partners into the play, including Mobil Oil. He drilled to below eighteen thousand feet. As at Pescadito dome, the temperature was so high that logging tools wouldn't work. The tools were frozen in a deep freeze, then sent back into the hole. They still wouldn't work. And there were no substantial shows of oil or gas.

The well was likely the first, and certainly the deepest, ever drilled solely in a search for a stratigraphic trap. Telegrams to Halbouty and his partners from other oilmen congratulating them on their boldness failed to temper the failure.

The failure that rankled the worst, however, was in Brazos County, in Central Texas. There had been no successful drilling in the county. To the north, however, were two fields, Fort Trinidad and Madisonville. Madisonville field produced its hydrocarbons from the Woodbine sand. It was not a good field. At Fort Trinidad, production had been found in the Glen Rose limestone, and it was this development that excited oilmen.

With some partners, Halbouty sought the Glen Rose limestone in Brazos County. He drilled to 11,500 feet on an anticlinal structure. There were no oil or gas shows in the Glen Rose.

But as the well was being drilled, the bit had passed through a three-foot layer of Woodbine, of which only eighteen inches looked like sand. Now, the Woodbine was reckoned to have thinned out this far south so as to be of little value as a reservoir. Nevertheless, Halbouty tested it. He perforated the casing in the Woodbine and nothing occurred. He acidized the formation as he had done years earlier at Pine Island field in Louisiana. No production. He "fractured" the formation. That didn't work. It was concluded that the thin Woodbine was too tight and too far south to produce. The well was abandoned.

Several years later Amalgamated Bonanza, a Canadian company, leased acreage in the area. If the company was seeking the Woodbine, it would have been geologically reasonable to drill to the north and east. The Canadians drilled to the south and east, a mile and a half from Halbouty's dry hole, and found a field to match the company name. The company drilled fifty-four consecutive producers from a Woodbine reservoir holding an estimated fifty million barrels of oil.

In newspaper oil pages and oil periodicals it was customary to locate a new well by saying it was such-and-such a distance from a known landmark. Almost all reports of Amalgamated Bonanza's triumphs said they were "1½ miles from M. Halbouty's dry hole."

Halbouty suffered. He suffered more when he heard that a remark was making the rounds—"Amalgamated made a well where Halbouty had a well and didn't know it."

He had done everything with his well that a prudent, responsible oilman should have done, but he didn't deny the remark or its implications. But he told a group of fun-pokers in the Houston Club: "I've made most of you fuckers rich by letting you follow me around. You wouldn't be eating and drinking in this club if I didn't have the guts to go out where the oil is."

They joshed him some more, but it was generally agreed that he had taken the hoorawing much better than had been expected.

To top things off, he went back to Alaska. By now, Halbouty's prophecy about Alaska was being fulfilled; Atlantic-Richfield had discovered a ten-billion-barrel reservoir on the North Slope at Prudhoe Bay. Halbouty teamed with Standard of California to drill on the west side of Cook Inlet—by now the wildcat side—and got a dry hole at twelve thousand feet. He was to go back later and get another duster. It was ironic that the man who had touted Alaska's potential at every oil forum, an independent who rushed in where the majors were cautious to tread, should be denied its riches.

These were the failures that bruised his spirit and slimmed his wallet. But during this time he ranged far and wide, drilling beyond the established hydrocarbon outposts, and in many instances he found the "annuity" oil he had found at

Pine Island field in Louisiana. So, while he was spending vast sums on the big failures, he was assuring himself of long-term income from more than a dozen small but steady productive fields. At his low point, he had income from twenty-eight oilfields in Texas, Louisiana, Wyoming, Mississippi, and Canada.

In his office Halbouty had a large filing cabinet that contained nothing but information on salt domes. He had been collecting the information for years. Much of the material had been written by him—scientific papers, magazine articles, and long notes he had written at random moments. And he had kept copies of papers written by others if he agreed with some or all of their conclusions.

As a Distinguished Lecturer for AAPG, he had delivered a paper on salt domes in the United States and abroad. Everywhere he spoke his listeners seemed to have an insatiable appetite for salt dome information. When he spoke before a large group of European earth scientists at The Hague, he completed his remarks in forty-five minutes. The Europeans kept him on the platform another two hours while they pelted him with questions. Then little groups began forming to debate various conclusions Halbouty had offered.

He resolved on his way home to write a book on salt domes. He knew there was no book on the subject in the world's literature. He also was aware that geologists fresh out of school shied away from salt domes because they had learned little of salt domes' complex structures. For these young people, his book could be a text.

"How in the hell are you going to find time to write a book like that?" James Clark asked him.

"I could never explain to a lazy man," Halbouty told his old friend and cussing companion.

He found the time. He wrote on Sundays and in the middle of the night. He wrote on his airplane between oilfields and speaking engagements. He completed the manuscript in eleven months. *Salt Domes—Gulf Region, United States & Mexico* was published by Gulf Publishing Company in 1967 to almost universal praise from his peers and the oil industry generally. An *Oil Daily* review suggested why.

"Halbouty has 'programmed' his book deliberately and carefully to appeal to a wide audience, one which includes not only the Gulf Coast geologists and geophysicists, but also earth scientists who work in any diapiric province in the world," the review said. "It is clear that he has familiarized himself with every aspect of his subject."

A. A. Meyerhoff, then editor of the *AAPG Bulletin,* wrote that the book was a "well of information that will never run dry . . . the first book of this scope ever published on the description of salt domes."

The book sold well. Halbouty received mail from every oil province. Most of it was congratulatory. Letters that pleased him were from earth scientists who said they had been helped by reading the book. Other writers queried him on specific points. He was never too busy to reply—and in detail. One correspondent maintained Halbouty was a "frustrated professor." Halbouty conceded that he would have made a "damned good teacher."

It was almost as thrilling as finding an oilfield. (In 1978 he was finding time and energy to bring out a new edition of the book with revisions and new material.)

His curiosity, his drive, his sense of impending doom, propelled him to Wyoming, Utah, and Colorado to study oil shale. Shale deposits in those states were believed to contain two *trillion* barrels of oil. The most important area was the Piceance Basin in northwestern Colorado, which included 1,380 square miles underlaid with deposits ten feet thick or greater that could yield twenty-five gallons or more of oil per ton of shale.

The oil shale in Colorado alone had been estimated as high as 1.5 trillion barrels, with the minable portion, which averaged twenty-five gallons per ton or more, containing 480 billion barrels of oil. Sixty percent, or 288 billion barrels, might be recovered with the technology current in 1968. Or it might not, Halbouty concluded.

Because much of the shale was on public land, Halbouty feared federal control of it. This time it appeared he had good reason to be frightened of the Washington ogre. And he was irritated with reports in trade magazines and in governmental bulletins of "trillions of barrels of oil" in the shale. And it particularly angered him that former Senator Paul Douglas of Illinois, a long-time Halbouty target, had been speaking out on the subject. Douglas had become an honorary chairman of the Public Resources Association, headquartered in Denver. His honorary co-chairman was Dr. John Kenneth Galbraith of Harvard University, the mention of whose name caused Halbouty to bridle. Both claimed that fabulous wealth belonging to the people would be given away if private enterprise developed the oil shale.

Halbouty wanted work begun on the oil shale immediately, and he said so, time after time. But he wanted to dispel the notion of "trillions of barrels," and he wanted the government to be what he considered reasonable.

In a "dear friend" letter, widely distributed on Public Resources Association stationery and signed by Douglas, was the assertion: "Every man, woman, and child of our country owns a share of the vast resources of recoverable oil, and the value to each in terms of total sales volume is at least $25,000."

Said Halbouty: "This sounds as if the shale oil already has been developed and sold, and the money set aside in a communal warehouse or deposited in a bank. All each of us has to do, Douglas implies, is write a check or get a shovel and withdraw our personal share.

"What nonsense! By similar reasoning, each of us could take an ax into our national forests, mark off our 'personal' trees, chop down our share, and put the lumber to our own uses."

Galbraith wanted the shale oil developed by the government to "protect the public interest, as opposed to the selfish and speculative interest" of private business.

"Again, what nonsense!" Halbouty roared. "Whatever private business does with oil shale accrues to the benefit of the public. The development certainly is not selfish. However, it is speculative; private industry could go bankrupt in this oil shale venture!"

Shale oil could not be refined by conventional means, he pointed out. It was deficient in hydrogen, and hydrogen would have to be manufactured and added during the refining process. Shale oil contained large amounts of such elements as sulphur and nitrogen, which would have to be removed.

"Nothing in the studies of shale oil operations suggests enormous profits, and shale oil operations are not going to offer any get-rich-quick opportunities. It is most peculiar that critics of the free enterprise system do not know this. They continue to believe the oil from shale will gush forth like the old oil gushers at Spindletop, Signal Hill, and El Dorado at 100,000 barrels per day without any expense."

Oil shale lands should be leased to private enterprise on a competitive bid basis, preferably by public auction, using a bonus system with a moderate, fixed royalty, Halbouty maintained. "The bonus system will result in the maximum leasing income to the federal treasury, will preclude political or other forms of favoritism, and will avoid speculation in leasing. Competitive pressures will insure that the lease bonuses reflect the profitability of shale oil operations. Therefore, the royalty should be held to a relatively low, fixed amount so that shale oil will not be at a cost disadvantage compared with competing energy sources."

He wanted leasing to begin at once so that research and development could quickly follow.

"Our current responsibilities are twofold: first, to accelerate by every possible method a daring, imaginative and greatly expanded program of exploration for conventional petroleum reserves" (he seldom left this out of *any* speech); "and second, to insist that shale oil lands be leased immediately to private industry for research and development to determine whether such shale oil can be produced economically and whether shale oil reserves can be included in estimates for our future energy requirements."

His first speeches about shale oil were in 1968. A decade later not one gallon of gasoline made from shale oil had been used to power an engine in an American automobile.

In private talks, Halbouty went much further in delineating the Middle East threat than he did in public speeches. In 1969 he was at lunch in Los Angeles with a friend and Lee J. Cobb, the actor, when he ventured that the Arabic states would some day attempt to dominate the earth.

Cobb was astounded. Israel's overwhelming victory in the Six Day War was only two years old. "Israel will win again if there's another war," Cobb said.

"Maybe," said Halbouty, "but it makes no difference. The United States will castrate Israel for them. When the Arabs get ready, they can shut off oil to this country or raise the price until it brings us to the edge of destruction. They'll use Israel as the excuse for what they choose to do."

He went on. "I'm thinking that some day the Arabs will nationalize their oil holdings, and they'll realize the power they have. The OPEC countries of the Middle East and Africa can become the economic center of the world, like this country is now. They can dictate who will survive and who won't."

"Are you suggesting total unity among the Arabs, a union of some kind?" Cobb asked. "They have different forms of government and, I suppose, different goals. It seems to me that they agree on only one thing, the destruction of Israel."

"That's now," Halbouty said. "The 'haves' will take care of the 'have not' Arabs. I think even Iran will join them. The power from oil will override the differences between them."

The discussion was continued at a table in the bar. Halbouty contended that the Arabs hated the United States for its continuing support of Israel, for years of thinking of Arabs as shiftless camel herders, for a thousand indignities real and imaginary.

"They remember when Islam was a powerful force," he said. "Their culture was dominant. When they get the economic power, they'll want it all back. It won't take long for them to think they *should* have it all back."

Cobb wanted to know what the Russians would be doing all this time.

"Enjoying themselves," Halbouty said.

"What about England?" Cobb asked.

"There may be fifty or more billion barrels of oil in the United Kingdom sector of the North Sea," Halbouty said. "I may be wrong, but that's what I think. The British won't have the worries we have. Hell, the British may be sending *us* oil for a change."

But what about that big discovery on the North Slope of Alaska? "Won't that help us?" Cobb asked.

Halbouty shook his head. "It will be years before that oil reaches the consumer in Minnesota or somewhere. There's a lot more oil in Alaska than has been discovered, but it will take years to do us any good. We're too far behind. Right now the majors are happy with things as they are. The government is happy. And the consumers don't know what the hell is going on, and don't give a damn."

Cobb had to leave; there was a script he had to read. He stood up and gave Halbouty his hand. He said gravely, "Do you honestly believe that we will see an Arabic union that will be able to impose its will on us and much of the rest of the world?"

Halbouty nodded, "I believe it, but I guess I'm nuts to be talking about Arabs dominating the world, or wanting to, when I can't get anyone to listen to me about their shutting off our oil supply some day."

Cobb thereafter would refer to Halbouty as the "unheeded prophet."

It was a good description, for two years later, in 1971, OPEC was able to raise its share of oil profits from 50 to 55 percent with a built-in inflationary factor of 2 1/2 percent.

And two years later the United States was faced with an oil "shortage" that no one seemed to be able to explain clearly. And later in that same year, while Saudi Arabia was talking about a 20 percent interest in the consortium exploiting its oil, Egypt and Syria attacked Israel with the support of other Arab countries.

And the United States, for all of its sins, was cut off from Arabian oil.

"Out of Gas," said signs in many a filling station driveway across America. In other filling stations customers were limited to ten gallons, and the lines of customers were long.

Natural gas, for reasons known only to oil and gas companies, suddenly was in short supply.

Prices for oil products and natural gas rose steeply.

OPEC kicked the price of oil to about twelve dollars a barrel.

It was a panicky time for Americans who listened to and read about dozens of plans to overcome the "energy crisis" while they watched their utility and gasoline bills increase.

The inevitable search for scapegoats produced several.

thirteen

AT THE HEIGHT of the energy crisis in December 1973, Halbouty was awakened in his San Francisco hotel room by a phone call from James Clark in Houston. Clark by now was the head of the Energy Research and Education Foundation at Rice University.

"For Christ's sake, it's only five o'clock," Halbouty snarled.

Clark was grumpy, too. "I don't give a damn, it's seven o'clock here. Reporters are trying to get hold of you, and when they can't, they call me. Do you want me to tell them where you are? They want to talk about the crisis."

"To hell with them," Halbouty said. "They know as much about it now as I do."

"First time I ever heard of you turning down an interview," Clark said.

"The same to you," said Halbouty.

The wire services had found Halbouty stories in their files, and stories of his predictions were sent across the country to newspapers and radio and television stations. Some of the story headings read:

GEOLOGIST'S WARNINGS
ARE NOW COMING TRUE

ARTICULATE INDEPENDENT
FORECAST ENERGY CRISIS

SPEECH 13 YEARS AGO
PREDICTED FUEL CRISIS

OIL CRISIS FORESEEN

HALBOUTY CAN NOW SAY:
"I TOLD YOU SO" IN '60

Ironically, while his warning speeches over the years had usually been reported on the business pages, the "update" stories most often were printed in the general news sections.

When Halbouty returned to Houston, reporters asked him about the energy crisis. "There's no energy crisis," Halbouty snapped. "It's a *stupidity* crisis. There's not a half a dozen men in Washington who know what the hell is going on, and nobody will listen to them."

What would he have the government do?

"The greatest incentive the government can give the oil industry is to leave it alone. And that goes for the wild-eyed environmentalists and conservationists. Listen! Right now we're fifteen years late in developing shale oil. We're a decade behind in coal liquefaction and gasification, and I don't know how many years behind in developing geothermal energy.

"Even worse," he went on, "thanks to the government and hysterical environmentalists, we are letting three hundred trillion cubic feet of gas stay locked in the hard formations of the Rocky Mountain states when it can be safely released through the application of nuclear stimulation—right now! Alaskan development is being held up at every turn. So is drilling on the Outer Continental Shelf." He threw up his hands. "So is every other Goddamned thing we should be doing."

He was reminded that restrictions on production in oil states had been removed, that oil and gas wells, his included, were running at full capacity for the first time since 1948. He was reminded also that domestic oil and gas prices had risen sharply. The implication was that Halbouty the independent oilman was doing all right.

"From the viewpoint of the independent, the shortage has been a boon," he said. "But only in that we're finally getting to produce some oil at a pretty good price. But look at the future for the country, for the people. What we need right now is more dry holes out in the boondocks.

"There are more than three hundred million acres of land under oil and gas lease in this country. All but a small fraction is held by the majors. This is ten times as much area as all of the productive oil and gas land in the nation. All of it must be considered prospective or someone is crazy for holding on to it to the exclusion of others who might think well enough of it to drill.

"That three hundred million acres under lease constitutes one of the greatest of all weapons against a supply shortage. We'll have to use all available technology and develop new, but the oil and gas is out there."

This challenge to the majors to drill or give up vast acreage to others who would drill was just about his last criticism of them. In the challenge was the implication that the majors had played a big role in the shortage by holding back domestic exploration and production in order to import cheap foreign oil.

Thereafter he joined ranks with them against the common

foes, Congressmen, bureaucrats, environmentalists, and anyone else with a harsh word for the industry. He laid the entire
blame for the gas shortage on "demagogues in Congress, pied
pipers in the Federal Power Commission, and pinheads in the
municipal gas and power companies." And he added sourly,
"The public went along with them."

In a speech for the Texas Daily Newspaper Association in
Fort Worth, it appeared as if he intended to repudiate Lee's
surrender at Appomattox. He had not approved, he said,
when first he saw the bumper stickers prevalent in Louisiana
and Texas—LET THE BASTARDS FREEZE IN THE DARK! Now
he wasn't so sure. The bumper stickers alluded to a peculiar
situation: Texas and Louisiana gas sold inside those states was
unregulated and therefore had no price ceiling. The same gas
sold elsewhere was regulated by the Federal Power Commission, which kept a regulatory lid on the price. As a result,
Texans and Louisianans paid considerably more for gas than
consumers in the north and east. At the same time, states
fronting the Atlantic seemed to be in no hurry to allow drilling off their coasts or the construction of refineries. Some Louisiana officials wanted to cut off gas shipments to them.

Halbouty didn't want anyone to freeze in the dark, "but let
us put the cause of the shortage and spiraling inflation where
it belongs—on the backs of the whimpering Yankees!" Their
whimpering prompted their politicians to keep gas prices
down until there was little exploring for the fuel, he said.

The expression was so cruel that it delighted editors, and
the speech was published widely in Texas and Louisiana with
"Whimpering Yankees" in the story headings.

In short order Halbouty attacked the federal government's
offshore leasing policy. Under the system, companies bid huge
sums for leases and the winning bids went into the United
States treasury.

Under his plan, Halbouty said, the successful cash bids
would not be paid to the government but would be an obliga-

tion on the companies to spend the sums in drilling and development. Once that amount was spent, the companies' obligation to the government would be fulfilled. If the full amount was not spent, the difference would be paid to the government in cash.

If a block of leases was dry, the difference between what was bid and what was spent on exploration and drilling would be paid to the government. "For example," he said, "a successful bid is fifty million dollars. Exploration and dry holes cost ten million dollars. At that point the company chooses to abandon the lease and, at the same time, pays the government forty million dollars."

He was disturbed, he said, by the surrender of 195 Outer Continental Shelf tracts after the drilling of only one dry hole.

Companies would be willing to drill more test wells under his plan, he contended.

The plan made enough sense to gain some adherents, but it got lost somewhere in Washington's mystifying energy maze.

The increase in domestic oil and gas prices increased Halbouty's income accordingly, but too many expensive dry holes and other expenditures had put him ten million dollars in debt by 1973. He had not always listened to his business adviser, O. O. Hare. He listened, however, when Hare told him he should sell some of his banks. "There has never been a time so favorable to sell," Hare said, "and there may never be again."

Halbouty had invested about five million dollars in the banks in the mid-fifties. They had made him money. Now he sold North Side Bank, the First National in San Angelo, and the First National in Paris for about thirteen million dollars. He kept the West Side National Bank he had built in San Angelo. His stock in the Bank of Texas, through a merger and other transactions, now was in Allied Bancshares, a holding company, and he retained it.

He paid off his debts.

Regardless of where Halbouty placed the blame for the energy crisis, the American public laid it at the door of the majors. The embargo had come as a shock, but no more than the sudden increases in prices for petroleum products and the staggering profits the companies made from the increases. No explanation the companies offered seemed to appease anyone.

Congress, in 1969, had lowered the depletion allowance from 27½ percent to 22 percent. Now a move began to abolish it completely.

Halbouty denounced the move with all of his vigor at every forum he could find. With typical hyperbole he said that elimination of the allowance would be "the greatest blunder Congress ever made." The move had come, he said, "at the very time when we need the allowance most." Indeed, he said, "the allowance should be increased!"

Nevertheless, Congress repealed the allowance for companies producing more than two thousand barrels of oil daily. For those producing two thousand barrels or less, the cut would come gradually, stopping at 15 percent in 1984.

The majors had not fought the measure. The allowance meant little to them in comparison with other tax advantages which most Congressman seemed to be unaware of or incapable of understanding. Independents had fought the measure because they relied heavily on the allowance in raising drilling funds from investors.

To Halbouty the action was "punitive." It was further proof to him that "radical Congressmen" would not be content until the oil industry was nationalized, or so he said.

Like other oil shortages in the past, this one disappeared shortly after the embargo was lifted in March 1974. A new glut was in the making, even though American consumers resumed their wasteful ways. Higher prices for gasoline and natural gas did not faze them.

President Richard Nixon, beset with Watergate, did little to decrease the country's dependence on foreign oil. Gerald Ford, his successor, did even less.

Instead of crumbling, OPEC grew stronger. As the glut grew, OPEC members cut back on production to keep prices high, just as any other cartel would do. And while Americans screamed at what they called the majors' "obscene profits," OPEC members took steps to skim the profits for themselves.

In time some Americans learned that OPEC was not just a group of Arab countries. Seven members were Arab countries—Algeria, Iraq, Kuwait, Libya, Qatar, Saudi Arabia, and the United Arab Emirates. But Ecuador, Gabon, Indonesia, Iran, Nigeria, and Venezuela also were members, and held great fields of oil and gas. They were friendly with the United States, but they joined other OPEC members in raising prices and maintaining them.

It was against this background that Congress presented President Ford with an energy bill, a "compromise package." It was a splitting headache for Ford. As a staunch right-winger, he had spent his political life supporting industry against attacks. He detested price controls and regulations, which he thought retarded industry, and he said so on many occasions. This package contained provisions for a continuation of oil and gas controls until 1979 and a rollback in domestic oil prices.

It was a bill Ford the House Minority Leader would have frowned upon. But President Ford, who had been given the office by Richard Nixon, wanted to win it on his own. He was running for the Republican nomination, and polls showed that Ronald Reagan was breathing in his ear. Primaries in the New England states were only months away, and his advisers had assured him that more people in New England bought oil and gas than produced it. It obviously was a time to let expediency rise above principle.

Most segments of the oil industry detested the bill and hoped Ford would veto it. Many in the industry thought he didn't understand it. Halbouty was in this number. He had met Ford when Halbouty was chairman of a fund-raising drive in Texas for Richard Nixon and Ford had come to Houston to speak for Nixon. Halbouty had introduced Ford and sat next to him through the two-hour meeting. He later confided in a friend: "Ford would make a damned good roughneck but never a driller."

Still, he was supporting Ford against Reagan, and he had contributed liberally to Ford's campaign. "What the hell," he told James Clark. "Give him a chance to win it on his own." Clark shook his head. "He's going to sign that damned bill, you wait and see. You're giving your support to the wrong guy. You're getting sentimental in your old age."

Senator John Tower, Republican from Texas and Ford confidant, called Halbouty. Ford's advisers wouldn't let him be caught with a spokesman from a major oil company who could plead the industry's position. An independent, however, was a different matter. Tower had taken an unofficial poll. Halbouty was the man to explain the industry's side to the President. Would he do it? Hell yes, said Halbouty.

In Tower's Washington office, Halbouty met with the Senator and three other Texans who were on hand to lend backup support—Peter O'Donnell, former Texas GOP chairman; Jake Hamon, a Dallas oilman, and Robert Stewart, a Dallas banker.

"We're getting only fifteen minutes of the President's time," Tower said. "You're going to have to lay it on the line, Mike. Do you need any props?"

"I know it all by heart," Halbouty said a bit grimly.

"I've guaranteed him that there are enough votes in the Senate to sustain a veto," Tower said. "It's up to you."

Ford greeted them in the Oval Office. He put an arm around Halbouty's shoulder and said, "Mike, it's so good to see you again." Halbouty responded warmly, but he acknowledged to himself that probably Tower had reminded the President that he had met Halbouty in Houston.

Also in the office was Frank Zarb, Federal Energy Administrator, and several of his aides. Zarb was pushing the President to sign the bill.

Tower said, "Mr. President, I've told you that I can guarantee that a veto will be sustained by the Senate. I want to repeat that guarantee. Now Mr. Halbouty will give you the facts about the bill."

Ford nodded for Halbouty to begin. According to witnesses, Halbouty did a fine job. As always, he was articulate. As always, he spoke with certitude. And he kept it simple.

The rollback in oil prices would decrease exploration for domestic oil, thus increasing the need for more and more imported oil. The rollback and maintenance of price ceilings would encourage consumption by keeping prices artificially low. The bill would permit OPEC prices to rise without permitting the domestic industry to take a single step in the nation's energy defense.

Ford interrupted Halbouty several times to ask questions. Finally he interrupted, saying, "I understood the independents were *for* this bill. You're telling me they're against it."

"They *are* against it," Halbouty said. "I can't imagine where you got the information that they're for it."

"Frank Zarb told me that."

"I suggest you ask him again, Mr. President," Halbouty said. But he didn't wait for Ford. He turned to Zarb. "What about it, Mr. Zarb?"

"Well," said Zarb, "the independent jobbers and independent marketers are for it."

"Certainly they are!" Halbouty said. "But you left the impression with the President that independent explorers and producers are for it, and that's not true!"

Ford told Halbouty to continue. Halbouty sucked in his temper and went on with his explanation. When he was through, he said, "This is just a compromise bill, Mr. President. It's not good. When you compromise, you're taking the bad and the good and putting them together and it still comes out a monster."

Tower asked O'Donnell if he had anything to add. O'Donnell said, "Just this: if the President signs the bill, I think he'll have a very difficult time winning Texas."

Stewart, the banker, said the effects of the bill would be disastrous financially. Hamon, the oilman, said: "I'll substantiate everything Mike Halbouty said about the bill."

The meeting broke up. Instead of the allocated fifteen minutes, it had lasted forty-five. Ford walked to the door with one arm on Halbouty's shoulder, the other on Tower's shoulder. "I've really learned a lot today," he said. "I'm going to do some real strong thinking about it."

Halbouty and the others went into the corridor, but Tower remained behind for a parting word with Ford. Halbouty pounced on Zarb. "Why don't you go back in there and tell that man to veto that bill?" he demanded. "You know it's the worst Goddamned bill you ever saw in your life! It's against the welfare of the people!"

"It was the best we could get out of Congress," Zarb said.

"Do you just have to *have* something? Do you want a trophy to hang on the wall that says you came up with *some* kind of energy package? That's all it is, a trophy to hang on the wall. You're just trying to protect your pride. Why don't you go in there and say, 'Mr. President, I made a mistake. The bill is no good.' He's right on the verge of saying he doesn't want to sign it!"

"I can't do it," Zarb said.

"You ought to. You're not protecting the welfare of the

people of the United States like you swore you'd do. I feel very bitter about it."

Tower joined the group. He pulled Halbouty aside, and the others drifted down the corridor. "I think we did some good, Mike," Tower said. "He says he sees this side of it now."

Halbouty flew back to Houston. He was pleased with what he had done. The next morning he received a phone call from Frank Ikard, president of the American Petroleum Institute. Ikard had attended a party in Washington the night before where Ford was present. The President had sought him out, Ikard told Halbouty, and talked about Halbouty's presentation.

"He said you really explained the situation to him, Mike," Ikard said. "He said you knew what you were talking about, and he said he understood more about the bill than he had before. I'll tell you, Mike, I think he's going to veto it!"

Three days later Ford signed the bill.

Halbouty told reporters: "I feel the President has caved in to the pressure of radical members of Congress, and I fear he has been misled by Frank Zarb to sign this legislative monstrosity for purely political reasons.

"As an American in search of leadership, and a Republican who has stood with his party through thick and thin, I am disgusted!"

To show his disgust, he called Ford's campaign headquarters and demanded that his contribution be returned. Then he called Ronald Reagan in California. "Governor," he said, "you've just got you a new supporter. Tell me what you want me to do, and I'll do it."

He stumped the state for Reagan and contributed generously to the campaign financially. Reagan handily took Texas from Ford. Halbouty was a delegate to the Republican National Convention in Kansas City where Ford won the nomination from Reagan.

Despite his disgust with Ford, Halbouty voted for him in the Presidential election; he could not abide Jimmy Carter.

The times were not all bad. In August 1974 Halbouty was general chairman of the first Circum-Pacific Energy and Mineral Resources Conference in Honolulu. It was sponsored by the AAPG and other organizations "to advance the exploration and development of the total energy and mineral resources of the Circum-Pacific area."

He was at his best as a mover and shaker at this kind of meeting where he could promote the free exchange of ideas, one of his articles of faith. The meeting was designed for representatives from countries bounding the Pacific, but countries removed from the ocean sent their people. To Halbouty's surprise, more than a thousand delegates from sixty-seven countries attended.

The conference was deemed a success. The Circum-Pacific Council, of which Halbouty was chairman, voted to hold a similar conference every four years. (The second conference was held in August 1978 in Honolulu, and attracted even more delegates and more discussion. Before it ended, Halbouty was deep in planning for the 1982 conference.)

A few months later, in February 1975, Halbouty received the Anthony F. Lucas Gold Medal from the Society of Petroleum Engineers. Earlier, in 1971, the Society had given him the DeGolyer Distinguished Service Award, but the Lucas medal particularly thrilled him. Lucas was one of his heroes. A mining engineer from Austria, Lucas had suffered one heartbreak after another before he brought in the great Spindletop gusher in 1901. It can be assumed that Halbouty identified with Lucas; the Austrian had studied salt domes in a search for sulphur and, after Spindletop, in a search for oil.

The citation with the medal said Halbouty was receiving the award "for his contributions to creative geology and petroleum engineering, new oil frontiers, total petroleum conservation, scientific literature, interdisciplinary communications between earth scientists and engineers, and public understanding of the industry."

He had done all of these things, certainly, and he had brought to each of them a passion and dedication few men ever muster. His peers who awarded him the medal knew him well. They knew all his foibles, and there were a dozen stories for each of them. But they respected him for his accomplishments, and they showed it with the Lucas award.

It almost seemed as if someone was listening when Halbouty complained about the three hundred million acres the majors held under oil and gas lease but had not drilled. Chevron approached him. The company had fifty-three thousand acres under lease in Beauregard and Allen parishes in Louisiana. Would Halbouty select a site and drill to twenty thousand feet for a 60 percent working interest?

Halbouty knew the area. It generally was considered too far north to be prospective for the younger horizons that produced in South Louisiana, and too far south to be prospective for the older horizons that produced in North Louisiana and North Texas.

Chevron had spent about $1.5 million assembling the leases and running geophysical surveys of them. The annual lease rental was $400,000.

Halbouty and Barber examined the area and Chevron's geophysical studies with a growing excitement. The seismograph had indicated deeply buried reefs in the Glen Rose and Edwards limestone formations. More, the formations had the same seismic characteristics as the great Golden Lane reef field in Mexico, though the Golden Lane reefs were at a much shallower depth.

In earlier years major oil companies had drilled deep dry

holes in North Texas in a search for hydrocarbons in the Glen Rose and Edwards. Later, in 1951, Mobil had drilled a dry hole to 18,651 feet—a formidable feat at the time—not far from the Chevron prospect. Halbouty and Chevron's scientists concluded Mobil had not drilled deeply enough.

A well to twenty thousand feet would cost Halbouty $1.2 million. He was aware that Tesoro Petroleum Corporation of San Antonio knew about the prospect, so he laid his information before that company's officials. Tesoro agreed to take half his action, or 30 percent of the deal.

There were 3,450 choice acres not in the Chevron block, which Halbouty and Tesoro had to lease on their own. Both wanted to drill on those acres, and Chevron agreed it was the best site.

Continental Oil Corporation wanted in on the deal. Halbouty dealt Continental half of his remaining 30 percent with Continental to pay two-thirds of Halbouty's costs.

The state of the drilling art had advanced to where deep drilling no longer was considered hazardous. The well was spudded, and the bit chewed down to 20,000 feet with ease— but it found nothing but some interesting geology.

All parties agreed to drill deeper. The bit was sent down to 23,472 feet. It found two good gas shows, one immediately below 20,000 feet, the other below 21,500 feet.

That marked the end of the happy days. A joint of oversized casing corkscrewed deep in the well, and the entire string of casing had to be pulled out. A very heavy mud was being used to hold back the formations at that depth. The extreme heat at the bottom of the hole caused the barite in the mixture to "settle out" of the mud into a concrete-like plug, which caught the drillpipe and froze it. Painstakingly, the drill pipe was fished from the hole.

Continental paid up and bowed out.

Halbouty and Tesoro continued trying to test the well. More months passed and more dollars were spent. Halbouty had to bow out. He did it reluctantly. Tesoro continued until two years had passed since the well was spudded in. The rig was "worn out" and so was a complete string of drillpipe. At one time the well briefly flowed some gas for Tesoro, but a conclusive test was never made. About six million dollars was spent on the well.

And 120 miles to the east, Chevron drilled down to 19,900 feet in Pointe Coupee Parish and discovered the False River field, one of the country's largest onshore gas fields. The production was not from the Glen Rose or Edwards but from the fabulous Woodbine sand just above them. False River was truly a giant, the kind of field Halbouty was saying would be found by drilling deeper and farther out in the boondocks.

Halbouty didn't give up on Beauregard Parish. He and Tesoro went back later on and renewed leases on 13,500 acres. In 1978 he was planning to drill there again, slamming a hard hand on his desk and exclaiming, "Damn it, there's a giant there! I know it!"

Halbouty was in Washington in June 1977 to receive the highest honor the AAPG could bestow—the Sidney Powers Memorial Medal, named for an early-day earth scientist who devoted much of his life to the building of AAPG.

This meant that Halbouty had received the three highest honors in both professional petroleum engineering and in professional geology—Honorary Membership, the DeGolyer Distinguished Service Medal, and the Anthony F. Lucas Gold Medal from the engineers; Honorary Membership, the Human Needs Award, and the Sidney Powers Memorial Medal from the geologists.

He was the only earth scientist to have been so singularly honored by the two great scientific and professional societies.

Halbouty's citationist for the Sidney Powers Memorial Medal was John D. Moody, formerly a Mobil senior vice president, now a geological consultant. Moody reviewed Halbouty's accomplishments and activities and concluded: "It is a mystery to me how this man finds the time and energy to do all of the things he manages to do, day in and day out. To know well such a genius and human dynamo is a real pleasure, and when these times are history, the name of Michel T. Halbouty will stand among those of the great geologists of all time."

fourteen

THE OIL & GAS JOURNAL of June 21, 1976, contained a story which began:

> Houston independent Michel T. Halbouty last week began work on the first of three rank wildcats in as many barren western U.S. basins.
>
> For years Halbouty has urged oilmen to drill in the relatively unexplored sedimentary basins of the western U.S.
>
> His ventures will be in the Payette basin of western Idaho, the Fallon basin of western Nevada, and the Harney basin of southeastern Oregon.
>
> All are non-productive basins. . . .

Halbouty had drilled many rank wildcats, but most of them had been located in what were considered oil provinces—areas where shows of oil or gas had been noted at some time, even though the shows had been detected at some distance from where Halbouty spudded in. Now he was really going out into the boondocks, as he had been urging the majors to do since 1960.

At a geological meeting in Denver he had blamed himself and other geologists for preferring to drill in known geologically proven areas instead of testing the unknown.

"Large reserves exist in Arizona, Idaho, Nevada, Oregon, Washington, and northern California," he had insisted. "They are lying right under our noses. To get them drilled requires the guts to shove the prospects down the throat of your boss or client."

He also had insisted that great reserves lay in what was called the Utah-Idaho-Wyoming "thrust belt," a swath of twisted, tortured formations paralleling the eastern edge of the Rocky Mountains.

He spudded in the James 1 in Payette Basin, Idaho, on June 19, 1976. The well was completed on September 21, 1976. It was drilled primarily to test the Horseshoe Bend formation, which was believed to be at considerable depth— about 12,500 feet—but Halbouty was eager to check some shallower formations as well.

A portion of Halbouty's written conclusions shows what occurred:

The #1 James penetrated approximately 2,270 feet of the Idaho (Pliocene) formation composed of alternating sands and shales. The Grassy Mountain Basalt (upper Miocene) was encountered at 2,270 feet and the Kern Basin formation (middle Miocene) interbedded with Columbia River and Owyhee Basalt flows penetrated to approximately 7,600 feet.

From 7,600 feet to total depth of 14,006 feet (1,500 feet below the projected 12,500 feet), the #1 James logged volcanic material, primarily tuffs, welded tuffs and basalts with occasional rhyolite and dacite of Sucker Creek (lower Miocene) age. At total depth, sediments of the Horseshoe Bend formation had not been reached, bottom hole temperature was 445 degrees, and drilling operations were discontinued. At 14,006 feet, the high bottom hole temperature indicated that the lower section had no commercial possibilities."

He had logged six oil and gas shows on the way down, between 4,100 feet and 8,120 feet, but none was considered of any significance.

The very next day after Halbouty plugged and abandoned the James 1, Amoco brought in the discovery well of the Ryckman field in the southwest corner of Wyoming, thirty miles north and eleven miles east of the Utah state line . . . in the thrust belt.

Ryckman was rapidly followed by a succession of fields in the thrust belt, among them Pineview, Yellow Creek, Whitney Canyon, South Chalk Creek, Dry Piney, Tip Top, and Hogsback. Other companies had wet their feet, and trade magazines were calling the thrust belt drilling the "hottest oil play in the lower forty-eight."

Halbouty, meantime, had plugged and abandoned his wildcat in Fallon Basin, Nevada, without seeing a single show of oil or gas. He was not optimistic about future chances in Fallon Basin.

He also struck out in Harney Basin, Oregon. His report read in part:

> Scheduled to drill to total depth of the 8,000 feet, the #1-10 Federal was drilled to 7,684 feet, at which point stuck drillpipe and excessive drilling costs resulting from innumerable lost circulation zones caused the operator to plug and abandon. . . .
>
> Information from the drilling of this well makes no contribution to a better understanding of the Mesozoic potential of the Harney Basin since no pre-volcanic section was reached.
>
> The Harney Basin remains a challenging and high potential petroleum province, and additional exploration in Oregon must be anticipated. . . .

That meant, of course, that some day he would be back. In the meantime, he praised Amoco for bringing in the Ryckman field. "You'd think," said a Houston geologist, "that he'd found the field himself, the way he goes on."

Halbouty had wept after his first wildcat failure in Alaska's

Kenai Peninsula. But almost twenty years had passed; that he did not cry now did not mean that the failures in the western basins were any less heartbreaking, but that he was, at long last, maturing. Not that he intended to accustom himself to defeat, but that he knew enough about his business now to accept that the odds were always against him. He had always known this; he had not always accepted it.

In a note to his secretary, Mary Stewart, he said: "Mary, don't fail to let anyone through who wants to talk about the frontier regions. Put them on the line. If I'm not here, Tom Barber will talk to them. He's as enthusiastic about it as I am. If we've learned anything out there, we want to share it."

John Moody, the citationist when Halbouty received the Sidney Powers Memorial Medal, had said, "It is a mystery to me how this man finds the time and energy to do all the things he manages to do, day in and day out. . . ."

Mary Stewart was one of the reasons Halbouty could do all he did. A tall, striking woman with alert black eyes, she worked almost as many hours as Halbouty did. She seemed to stay one step ahead of his business, keeping his voluminous records straight, fending off those who would take his time needlessly, making suggestions to him calmly during crises, ignoring his bad office manners all the while. If anyone was indispensable to Halbouty, it was Mary Stewart.

She saw to almost every detail of his operations, including finding space on the office walls for the multitude of plaques representing the honors bestowed on Halbouty. Halbouty also loved pictures, and these too graced his reception office walls. "You'd think he was dead and this was a shrine," said a sour-lipped visitor.

She had to find space for a special plaque and a special picture in 1977. Through the years Halbouty had made major contributions to his old school, Texas A&M University, through services and financial assistance. In 1968 the school had presented him with the Distinguished Alumni Award, the institution's highest honor for an individual. Now the school named its geoscience building "The Michel T. Halbouty Geoscience Building." A picture of Halbouty standing by one of the nameplates on the structure shows a man few people ever met: an emotional man near tears, his strong face softened like that of a man in love, a man willing to bend a little—or perhaps a lot. God save us, there is a hint of humility in that softness.

And there was the William T. Pecora Award, given him by the National Aeronautical and Space Administration and the Department of the Interior. The award also honored its namesake: the late Dr. Pecora, as director of the United States Geological Survey and Undersecretary of the Interior Department, saw the possibilities of using satellites to study the earth's resources from space. Halbouty, one of Pecora's closest friends, was right behind him, and he became a constant and effective advocate of such a program.

He spoke on the subject at many meetings. He visited universities and intrigued imaginative geoscience students with "remote sensing" possibilities. He wrote papers about it. One paper, "Application of Landsat Imagery to Petroleum and Mineral Exploration," delivered at the 1976 annual AAPG meeting, was considered a classic in the field.

The Pecora award said in part, "With his proven professional judgment and his forceful and convincing advocacy of the earth satellite program, he has done more to win its acceptance and use, especially in petroleum and mineral exploration, than any other individual."

But there is no doubt that his alma mater's naming the geoscience building for him was the high point of his later years. He identified with Texas A&M like some men do with established families or corporations to which they have given many years of service. He had little need of anchors, but the school was the rock to which he tied his sometimes tattered rope. He went there often to lecture, gaining as much as he gave.

The graduate scholarship program Halbouty began in the lean days of 1946 provided funds for fifty-six students through 1978. Strangely for him, Halbouty did not show a proprietary attitude about these students. They were never his "boys." But almost weekly a letter from a graduate comes into his office, some from those who have been in the field since the forties and fifties. In most instances they have read something about Halbouty in some trade magazine and simply want to acknowledge that and say thanks again. He replies to all of them. His notes are stilted; they show none of the great pride he feels in the graduates' accomplishments or his assistance. It is a very private contribution.

Halbouty read the letter a second time. He placed it carefully on his desk and looked up at Mary Stewart. "Do you believe this?" he asked.

She laughed. "Not with the potshots you've taken at them."

The letter was on the stationery of the Organization of the Petroleum Exporting Countries—OPEC. It said:

> The Secretariat of the Organization of the Petroleum Exporting Countries is contemplating convening a Seminar on the "Present and Future Roles of National Oil Companies" (with particular emphasis on the National Oil Companies of its Member Countries).
>
> It is envisaged that this Seminar will commence on Monday, 10th October, 1977, and last for three days. The participants will be drawn from the National Oil Companies of the OPEC Member Countries, as well as from these countries' Oil Ministries. The number of such participants is foreseen as approximately 100 to 120 persons.
>
> The OPEC Secretariat would greatly appreciate the opportunity of this Seminar to seek your valuable experience, and has

pleasure in inviting you to deliver a paper with the title "Future Trends in Exploration to Discover New Hydrocarbon Reserves throughout the World and the Role to be Played by OPEC Member Countries' National Oil Companies," or any other which may be related to the subject matter.

In the event you accept this invitation, would you kindly inform the OPEC Secretariat as soon as possible.

We look forward to receiving your affirmative reply, and meanwhile will keep you informed of other developments as they occur.

The letter was signed by Ali M. Jaidah, OPEC Secretary General.

Halbouty got up from his desk and went to a window overlooking busy Westheimer Road. More than twenty years earlier he had erected his first building across the road and watched the daylilies bloom. He had been far out in the country then, but now the rich and sprawling city reached for miles in every direction from him, and more than two million people were his neighbors.

Mary Stewart stood and watched the brooding figure at the window.

"By God, Mary," Halbouty said, then lapsed back into silence.

Finally Stewart left the office and went to her desk. "I don't know what he was thinking," she said later, "but I was thinking that the letter meant he had reached the absolute top in his profession. Here was the most powerful group on earth—I mean non-military—asking for his advice in his field. He was shaken by the invitation, that was obvious, but I knew nothing on earth could keep him from going."

He went. He had been to Vienna before, on vacation and on business, but that had been prior to OPEC's ascension and the selection of Vienna for its headquarters. He loved the ancient city for its gingerbread architecture, but he stayed at the Hilton.

He had hardly closed the door behind him in his suite when the phone rang. An OPEC representative wanted to visit him briefly. Come on up, Halbouty told him.

The squat, swarthy man was apologetic. Halbouty must not mind the guards, he said. Past kidnappings and bomb scares made them necessary. There was nothing to be alarmed about, but still . . .

Halbouty hadn't noticed any guards. His visitor assured him there were guards about. Halbouty would register for the seminar in the hotel. He would be photographed, and he would wear the photograph as a badge on his lapel at all times. He would be taken to OPEC headquarters by bus and be returned the same way. The visitor bowed and left.

Halbouty was processed the next morning. As he walked to the bus he saw men with carbines casually guarding him and the others who moved into the vehicle. Guards rode with him on the bus and others followed the bus in a black Mercedes.

The OPEC headquarters building was an anachronism, a modern chunk of gray limestone dropped among the domes and spires of Obere Donaustrasse. Halbouty was questioned twice, once as he entered the building, again in the large vestibule. After that he was treated royally. Every person who approached him appeared to be a host, each trying to outdo the other in courtesy and geniality.

He had been told that from 100 to 120 persons would be present. He guessed the number at closer to 300; there was not an empty seat as the seminar opened. He was told that newspersons from thirty-eight countries were in attendance. It seemed as if every important oil company had sent at least one representative to listen and take notes. Black men, brown men, and white men were there, men from oil-producing countries and men from oil-hungry countries. All were there at OPEC's special invitation.

Halbouty was to be the first of fourteen speakers over the three days. Chairman of the meeting was Dr. Mohammad

Sadli, Minister of Mines, Indonesia, but the man who held the reins was Ali Jaidah, the Secretary General of OPEC and Qatar's leading oilman.

What Jaidah said that day, and what Halbouty said, was printed in part in newspapers around the world, while the remainder of the meeting was largely ignored. Halbouty had not intended for it to be that way. He had written his speech with care. He had cautioned himself time and again not to rise to any bait that could lead to controversy. And he followed his game plan.

But Jaidah, in his welcoming speech, was aggressive and critical of the western oil industry . . . and Halbouty's prepared speech, which he delivered without a single deviation, seemed to answer Jaidah's challenge.

Jaidah knew that Halbouty had not "answered" him, and so did many others present. But the reporters apparently didn't, and neither, apparently, did some of the Arab delegates.

Jaidah spoke beyond the meeting; his message was for western governments and board chairpersons of multinational oil companies; its thrust: We're tired of simply producing oil, we want to be a major part of the whole shebang.

It was always OPEC's aim, he said, to take over the decision-making function in the setting of oil prices, and this had been done. Now OPEC wanted to gain complete control of the hydrocarbon industries in its sovereign territories. But the drive for overall economic development had not been pleasant.

We have been buying technology at *exorbitant* prices, whilst turn-key projects have proved to be tied to the suppliers of technology for patents, spare parts, operations, research, and so forth.

The present terms of transfer of technology are a source of deep concern to us, not only because they are of a *grudging* nature, but also because we are denied access to the markets for the products in the developed countries. The argument that the ad-

vanced consuming countries have surplus refining and petrochemical capacity is totally unacceptable to us because, every
day, we hear of new plants being constructed in those very countries. . . .

How can we talk seriously of cooperation when OPEC Member Countries, taken together, command a meager 6 percent of
the world's refining capacity? And what about our negligible
share of the world's petrochemical industry? A mere 3.2 percent!
Neither can one help pondering about the significance of the fact
that less than 3 percent of our Member Countries' crude oil
exports are transported by means of our own tankers."

And here Jaidah delivered his threat.

> *Gentlemen, I must say, in all seriousness, that unless greater progress is
> made in redressing the imbalances in this area, our Member Countries will
> have no recourse but to adopt collective strategies to achieve their aims.*

Jaidah didn't say what the collective strategies would be,
but everyone present—as well as the rest of the world—was
acutely aware of the 1973 embargo and the subsequent qua-
drupling of crude oil prices.

A person less sure of himself than Halbouty might well
have wondered what the hell he was doing there. He was a
wildcatter, a true independent. He owned not a share in
OPEC oil at any stage from drilling to marketing. Since his
arrival he had been assured several times that he was welcome because the OPEC nations wanted to know where the
new hydrocarbons would be found and how they would be
found—and that they thought he was the man to tell them.

He told them. He ranged the world, telling them of the
basins, near and remote, that he thought contained hydrocarbons and why he thought so. It was a brilliant, encompassing
survey. (A Scottish reporter remarked, "He didn't leave out a
square inch of the universe, did he?")

In the past Halbouty's speeches had pleaded that the
United States be drilled in every likely corner to reduce dependence on foreign oil, Arabian and African oil particularly.

Now he pleaded for drilling in every nook and cranny of the globe for the betterment of the world's people.

No one present could have failed to be impressed by the extent of his geological knowledge of the globe. He spoke of the geology and potential of the Soviet Union and other Iron Curtain states as familiarly as he spoke about Texas, Alaska and, yes, the Middle East. Several times before his speech he had been told that he had been invited to the meeting primarily because of his expertise in this regard.

But his basic theme was a familiar one, threaded through a much bigger fabric. The world was in an oil glut, but it would not last forever. Global demand would rise despite conservation in some countries. Exploration was not keeping pace.

> Finding as much oil and gas as possible stands out as a very high global priority. All of us, whether in the government sector or the private sector, whether in a market-controlled economy or a centrally planned economy, must do our utmost to accelerate the discovery of new oil and gas in order to provide a smooth transition into the post-oil era. If we fail, a world catastrophe is a not improbable outcome. All of us will be losers.

But what was occurring?

> The picture that emerges is not a pleasant one. Only in North America do we see any response to the need for increased exploration. But the bulk of this effort is unfortunately directed toward small fields, as most of the large fields have already been found. Outside of North America exploratory drilling has remained flat. . . .

Geophysical crews were not as busy as they should be, and private and national companies were not leasing or taking over acreage for examination.

> Again, we find that the worldwide petroleum industry is not responding to the need for accelerating exploration. So, philosophically, those of you who are associated with government com-

panies should ask yourselves these questions: Are you involved in the solution to the well-perceived need for increasing exploration—or are you a part of the problem? Are you promoting exploration within your area of responsibility—or are you inhibiting it?

Are you principally concerned in establishing the integrity of your operation—or are you trying to find more oil? If you are candid with yourselves, the answers will be a mixed bag, and—yes, I'll answer for you—you *could* do more exploration, perhaps much more.

The rate of discovery of Middle East oil had dropped off markedly since the middle 1960s, he reminded his listeners.

And, as usual, he took on governments wholesale, regardless of political coloration, blaming them to a great degree for the lack of exploration. And at this point he delivered the remarks that the news media interpreted as an "answer" to Ali Jaidah.

I would caution that governments, as well as their companies, should not overly restrict private company profits through exaggerated fears of exploitation. It is evident to me that the private companies must still play the major role in securing new petroleum supplies. Without them, I doubt that a world-wide successful effort would result.

Your philosophy may be to hang tough in the hope that the private companies will eventually come to heel, but I wouldn't bet on it if I were you.

Perhaps government oil companies on their own, given sufficient time, could find the major resources believed to exist. But the whole world, collectively, does not have the time to wait for either the transfer or development of experience and expertise.

The private companies have in hand, and are continually developing, the scientific and technological know-how and long-term experience to assist in achieving our desired goals in the shortest possible time.

I strongly emphasize that there should be coordinated and effective joint ventures between government companies and private companies in the exploration, development and production effort—and by this I mean a true, a real true profitable partnership—for the new petroleum supplies of the future.

There is absolutely no question in my mind that an alignment of government and private companies and a unified cooperative exchange of science, technology, know-how and experience will be more effective and profitable for the government companies.

However, this joint effort should be done on an equitable basis where the private companies are given a respectable share of the production and the profits therefrom.

Thus, it will enable the private companies, either on their own or with appropriate government companies, to explore the many new high-cost, high-risk frontier areas that must be explored.

During the break after his presentation, Halbouty was besieged by representatives from both Arab and African countries and the major oil companies. Arabs and Africans were trying to question him; western oilmen were trying to congratulate him in the mistaken belief that he had deliberately pleaded their cause. Halbouty kept emphasizing to both groups that he simply had said what he believed would be the best for the world's people. Private and government companies would have to cooperate if the hydrocarbons were to be found.

"What kind of formula would you suggest?" an Arabian asked.

Halbouty knew the man wanted a mathematical formula, and he knew he wasn't going to offer one. "Trust and cooperation," he said. "It's very simple."

The man smiled.

"What makes you think we need the private companies?" another man asked.

"I thought I made that clear in my presentation," Halbouty said. "To be a bit more down to earth with you, you just don't know how to get things done without them."

"But you say trust the oil companies," another man said. "How can we trust them when they stole us blind for years? Where does the trust lie? They came in here and gave us nothing and took almost everything."

Dr. William Brown, a European geologist associated with

Robertson Research, Ltd. of Wales, told an interviewer: "I was especially interested in hearing Halbouty's response to that question. He maintained his poise admirably, I thought. He said, 'I don't buy that talk, that the companies came in and stole you blind. You wanted them in your countries, and when they came in they asked you what you wanted in return for concessions. You told them, and you got it.' "

Dr. Brown smiled. "Halbouty moved his hands so that it looked as if he were drilling a hole. He said, 'When the companies came in, you didn't even have an earth auger.' Some of the OPEC members laughed at that. But Halbouty continued. He said, 'You gave the companies long-term concessions, but when they found the oil for you, you kicked them out. Now *you* tell *me* where the trust lies.'

"Then Halbouty held up his hands. He said, 'No more talk about who did what to who. You and the companies need each other, and you ought to accept it. Both of you ought to accept it, and get out of here and find the oil the world needs. And you can start right in the OPEC countries. There's a lot of oil still hidden there if you and the private sector just get out and hunt for it.' "

Dr. Brown added: "I heard at least a dozen people say—OPEC members and oilmen alike—that Halbouty's presentation cleared the air with a vengeance, and revived a dialogue that had been a monologue since the embargo and the big price hike. It so happens that I agree."

A few of the speakers who followed Halbouty seemed in general agreement with him—more oil must be found and produced whether by private companies, national companies or in joint ventures between them. Let's bury the hatchet, these speakers seemed to be saying, and get on with the business at hand.

Others, however, could not conceal their bitterness toward the multinationals, even when the thrust of their presentations was strictly parochial.

Halbouty had talked at length about exploring for and finding oil in harsh environments. In every case of recent oilfield discoveries he mentioned, the find had been made by a private company. His implication, intended or not, was that the private companies alone were willing to seek the high-cost crude, that national companies hadn't the know-how and fortitude for such adventures.

But one speaker, Nordine Ait-Laoussine of Algeria, pointed out that sixty billion dollars would have to be mobilized to produce the equivalent of 1.5 million barrels of oil daily from U.S. and Canadian shale oil, tar sands and transformed coal; twenty billion dollars to develop Alaskan oil reserves; ten billion dollars just to move 2.5 billion cubic feet of gas per day from the Canadian Arctic to where it is needed.

In a single day, he said, OPEC countries destroyed enough gas by flaring to equal one-and-a-half times the anticipated 1990 production from tar sands, shale oil, and coal liquids combined. The world had only a hint, he said, of OPEC gas reserves. And in addition to OPEC's tremendous known oil reserves, about 750 billion barrels were waiting to be discovered in the Middle East and North Africa alone.

Ait-Laoussine was plugging for higher prices for OPEC oil and gas based on the long-term supply cost of alternative energy sources and the cost of oil found and produced in frontier areas. Oil prices, he said, must increase *now* by 50 percent if the development of alternative sources was to be encouraged.

He went back to Halbouty. "The private companies, and especially the multinational ones, have the capability to discover sizable amounts of additional reserves. In this regard, the OPEC countries are still a good prospect; most of them are in the 'fertile crescent' identified by Mr. Halbouty." His implication was that he welcomed multinationals, under strict terms of a new kind of partnership, to drill in the OPEC countries.

Halbouty wanted nothing more. He only wanted energy for the world's people.

Ait-Laoussine, however, said he was puzzled by Halbouty's formula of trust and compatibility. He was not long on trust, Ait-Laoussine implied, based on past experience.

But he said:

> The most positive thing which could result from this seminar is the acceptance of the need to anticipate *now* the pending crisis of the next decade. If we can get together to share the responsibility of getting this message across, this seminar will have achieved much. If we can go further, and agree that the most effective way to encourage investment, not only in energy production but also in energy saving, is to manage energy prices by a planned move towards a long-term replacement cost parity, we will have taken a significant step towards solving the problem.
>
> It is essential that consumers everywhere cease to look on higher oil prices as a short-term burden and start to regard them as an insurance premium against scarcity.
>
> If the mistakes of 1973/74 are to be avoided, sterile talk of dependence must be forgotten, and a new sense of energy inter-dependence, in a spirit of cooperation, created.

Halbouty did not fault Ait-Laoussine in most of what the Algerian said. He believed the "fertile crescent" contained vast unproved reserves that could be found and produced much cheaper than the oil in the so-called "harsh environments." He hoped that the OPEC leaders would ask the multinationals back into their countries in deals fair to both sides because he thought the national companies couldn't find and produce the oil alone.

But he also believed that the big money should be spent to get the oil from the harsh areas because every drop would be needed. And he thought the national companies should participate in this effort. He couldn't blame the national companies for wanting to enter into all phases of the industry, but the first priority was to find the oil wherever it was.

"I think Ait-Laoussine is as genuinely concerned about all the people in the world as I am," he told some at the meeting. "I hope so, and I think I heard it in his presentation, between the lines a lot of it, and I read it in his face. A man's face means a lot to me about the way he believes."

He left Vienna satisfied that he had done his job. He had told his listeners where the oil was and how it could be obtained, and they had accorded him strict attention and, apparently, belief. He flew back to Houston with a feeling of well-being and with compliments ringing in his ears.

Many months later Halbouty was attending an international meeting of geoscientists. Someone in a small gathering asked him to pinpoint where he thought giant fields might be found around the globe. He did so.

In the group was an Iranian economist who had attended the OPEC seminar. The news media was full of accounts of terrible unrest in his country. When Halbouty concluded his extemporaneous remarks, the Iranian said, "Ah, Mr. Halbouty, you talk like our geologists. They keep saying there is much oil in Iran to be discovered. They are never silent on the subject."

"You ought to listen to them," Halbouty said.

The Iranian laughed. "Oh no. In my country the geologists recommend but the economists make the decisions."

"Yes," said Halbouty, "and that's why your country is so all fucked up."

fifteen

A son had been born to Linda Fay and her husband, Dennis Hewitt, in 1975, and the couple produced a daughter, Meghann Fay, in September 1978. Naturally enough, the son was named Michel—Michel Dennis Hewitt. A picture of him in Halbouty's office shows the boy peering through a knothole in a tall fence. It is a clue, perhaps, that he inherited his grandfather's curiosity.

Tom Kelly, in 1978, was running for governor of Alaska as an independent. While heading up Halasko's Anchorage office, Kelly had become a staunch friend of Walter Hickel. He had returned to Houston as general manager of Halbouty's interests, but had gone back to Alaska when Hickel was elected governor in 1966. Hickel appointed him Commissioner of Natural Resources. In that position, Kelly had responsibility for Alaska's parks, forests, land, water, minerals, agriculture, and hydrocarbons. It was he who arranged for the Prudhoe Bay oil sale for the state, which allowed the exploitation of the North Slope. In later years he managed his own geological consulting firm.

Hickel also was running for governor again, but was defeated in the Republican primary.

Halbouty was deeply involved in Kelly's campaign, and Mary Stewart's office looked like a campaign headquarters. She was ramrodding an "Aggies for Kelly" program that was aimed at extracting dollars from Aggie pockets.

(Kelly was a loser in the November 7 general election, trailing the Republican and Democratic contenders. Hickel, who had jumped back into the race as a write-in candidate after his primary defeat, also was an also-ran.)

Kelly had married Jane Moore of Wichita Falls and in 1978 they were the parents of Mark, twenty-one; Brooke, fifteen, and nine-year-old twin daughters, Adrienne and Isla.

Halbouty was no doting grandfather. Just as he was not an attending husband, he saw the grandchildren at his convenience. Fay, with homes in Houston, Palm Desert, and Colorado Springs, spent much of her life without Halbouty. He simply was on the go most of the time. Age did not diminish his restless energy; he had to be there, he had to be yonder, learning, teaching, hunting giants in the earth. Fay had known this about him from the beginning, and she had built her own life around this knowledge. She had been angry with him when he had been gone for months in his great crusade against the Texas Railroad Commission, and after that she apparently had not made her peace with Halbouty's constant movements.

He had grown handsomer with the years. Despite his conceits—large and small—he had become more charming. He had that property, never completely identified, of walking silently into a crowded room and commanding instant attention. He looked younger than men ten years his junior, and he acted like it. There was an aura of masculinity about him that was attractive to both sexes.

If love had died over the years, if friendship had been blurred by Halbouty's absences and involvements, Fay's pride in his accomplishments had dimmed not a whit.

This was never more evident than on the night of September 27, 1978, when more than eleven hundred persons gathered in the Imperial Ballroom of the Hyatt-Regency Hotel in Houston to honor Halbouty. It was the largest crowd ever assembled in that ballroom, a hotel official told the appreciative audience. Halbouty, upon being introduced, said, "Fay, stand up and let these people see how pretty you are." And she was pretty, her pride in Halbouty shining on her face.

This was a big night for the wildcatter. He was presented with the City of Hope's Spirit of Life Award. This is one of those awards the wealthy of the country pass around among themselves to raise money for worthy causes. Those at the Halbouty affair paid a hundred dollars a plate for a fine meal to help finance medical research, something Halbouty had been pouring money into since long before he achieved millionaire status. Still, it was different from the ordinary run of such affairs.

For one thing, Halbouty had given to medical projects when he couldn't afford it, gambling that he would be able to cover all bets—and winning.

For another, most of the people at the affair knew him, and were there because they wanted to be there.

And for yet another, men Halbouty respected stood up and praised him. They had known him for a long time, were aware of his strengths and weaknesses, and their affection for him glowed in every word they uttered.

One was Bill Clements, an oil drilling company executive and Republican candidate for governor of Texas. Clements said what could be expected of a Republican candidate talking about a Republican financial contributor. But in the overtones of his speech was a warmth generated by an old and steady friendship. (Clements won his campaign, becoming the state's first Republican governor since 1874.)

Another was Robert O. Anderson, board-chairman of Atlantic-Richfield. To Anderson, Halbouty was a wildcatter, "the premier wildcatter in this country," he said, "not only in the oil patch but in ideas, with the unique ability to transform dreams into reality, an individual who combines a strong sense of business with an equally strong compassion for humanitarian causes . . . a wildcatter in every fine sense of the word."

The last was Leon Jaworski, the famed Watergate prosecutor who, as a young lawyer, had defended Glenn McCarthy in the multitude of lawsuits that grew out of the West Beaumont blowout in the mid-thirties. Halbouty, Jaworski said, was his expert witness. "Mike literally took me by the hand to teach me the basics of oil-well drilling—scientific facts about sand formations, gas pressures and mechanical features, about oil-well equipment, including blowout preventers and master-gate valves.

"We suffered together through the preparation as well as the trial because it was a tough case as lawsuits go. To me the experience was worth all of the stress and agonies it entailed, for from it emerged a friendship fast and enduring—one I view with pride today.

"Incidentally, we won the lawsuit. You should have known this because lawyers never bring up lawsuits they lose. . . ."

Jaworski said he felt he should mention "that Mike can run the scale from a mellow voice to a siren-like scream faster than anyone I have ever known." But, Jaworski said, "he laughs easily and oftentimes at himself."

In presenting the award to Halbouty, Jaworski said, "Well, Mike, I've bragged on you long enough. It should be sufficient to justify the retainer fee to my law firm for another year."

Halbouty, too, recalled that lawsuit of more than forty

years earlier. "In those days, Leon, both of us were young, naive and unspoiled—and we had solid black hair. Now, look at us. We're just a couple of white-haired, sophisticated, spoiled old codgers." ("Old, my foot!" a woman whispered to her companion. "They're probably the two youngest men in the house!")

Halbouty had been told that he could speak on any subject he chose. On that very day the U.S. Senate had voted, at long last, on the natural gas provisions of President Carter's "energy package." The Senators had voted to phase out government regulations on gas rather than to deregulate immediately.

Reporters, knowing that Halbouty considered immediate deregulation as vital to the country's economy and security, had attended the party to hear him lambaste the Senate and the President, and thus provide them with a follow-up story. They were disappointed. Halbouty directed his speech to *all* government regulations, all bureaucracy. It would have been a standard right-wing attack on "creeping socialism," but Halbouty's phraseology and passion made it sound for all the world like a call to arms.

"The tyranny of government is everywhere!" he declared. "There are no mass executions of people, but there is a mass execution of the people's rights! We no longer live in a nation controlled by the principles upon which it was founded. Instead, we live in a land where federal politics dominates, controls, tyrannizes. The tentacles of a monstrous bureaucracy are enveloping and crushing our incentives and enthusiasm to produce and prosper—which breaks the morale and spirit of man."

Then he lifted the partisan audience to its feet with these words: "I would as soon be governed with a rifle at my head as to be bound hand and foot and gagged with the red tape of regulation!"

It is doubtful that all of his applauders actually were willing to trade red tape for the rifle, but in that electric moment they were willing to suspend rationality. And it was certain that most of them could well believe that Halbouty meant exactly what he said. As Jaworski had said earlier, "There is no equivocation, there is no ambivalence in what he says."

Several days after the event, Mary Stewart was shuffling through a stack of telegrams from persons who had been invited but had not gone. Halbouty had not yet read them. They were the usual run of such messages, and she was scanning them without great interest. Then she came upon a brief letter from a Texas oilman. He had been at the party and wanted Halbouty to know that he shared Halbouty's joy. The relevant part said:

> I couldn't help but reminisce back twenty-two years ago, when I went flat broke, and I was in your office talking about a deal. As I was preparing to leave, you said, "Do you have *any* money?" I said, "No." You cursed me for being so stupid, then picked up the phone and told your secretary to write me a check for $5,000. I can't tell you what this meant to me at the time. Not only was the material help appreciated but, more important, it lifted me out of a bad state of depression. Thanks again, Mike!"

The next day Mary Stewart saw Halbouty thumbing through the messages. She pointed out the letter to him. He read it. "I'll be damned," he said softly. "I had forgotten all about that."

"Did he ever pay back the five thousand?" Stewart asked.

Halbouty lifted his thick brows. "Hell, I don't know. I suppose so."

He picked up several other awards during 1978, including the Horatio Alger Award, which he went to New York to accept. Mary Stewart had to find space in the reception office for these additional plaques, and a building maintenance worker, watching her juggling the mementos, muttered, "She'll be hanging them from the ceiling before the year is out."

Halbouty seemed unaware that a visitor might consider the display ostentatious. There is no record of anyone's ever telling him so, in any event.

Howard Warren, a wildcatter with whom Halbouty had drilled into the Pescadito dome, and who was still wildcatting though in his seventies, told me: "I'm sure nobody deserves these awards any more than Mike, but I wish he'd quit winning them and quit making all these speeches and get his boots muddy in the oil patch again. I miss the rascal."

I pointed out to Warren that Halbouty was involved in drilling in five widely scattered areas at that very moment.

"It's not the same," Warren said. "He's got other people doing what he ought to be doing. Or that's what I think."

I told Halbouty what Warren had said. "He seemed sort of sad, Mike."

Halbouty was pensive for a moment. He nodded. "He's right. I ought to be out there."

Within an hour, however, he was deep in plans for a seminar to promote more and better use of remote sensing in finding hydrocarbons. And he was making notes for a paper he had agreed to deliver at an international meeting in Bucharest. And he was raising money and sparking ideas he hoped would help Ronald Reagan win the Republican nomination for President.

Nevertheless, he still kept his hand in the search for oil and gas. In October 1978 he was participating in drilling in Kansas, Wyoming, Colorado, and Hardeman County, Texas.

And, for the first time, he was trying his hand offshore. Halasko held a 7 percent interest in 97,920 acres of leases in the waters of Lower Cook Inlet. Atlantic-Richfield and Chevron held the remaining 93 percent. The acreage had been won in bidding during the Lower Cook Inlet Sale of late 1977. It had cost the companies $50,543,492.68.

It was certain that Atlantic-Richfield and Chevron did not need Halasko's participation to finance the deal or to drill the large block. It was a reasonable assumption that Halasko was invited in because of past associations and the regard in which Halbouty was held. In any event, no one in Atlantic-Richfield or Chevron could have been more eager to begin drilling than Halbouty.

Like a good omen, he received a phone call from Tom Kelly in Anchorage on October 25. "We've tied her in, Mike, and everything's working and she's flowing good."

Halbouty called in his chief petroleum engineer, Doyle McClennen, to share the good news with him. The lonely gas well on Kenai Peninsula—the only successful well he had drilled in Alaska—had been tied into a line belonging to the Alaska Pipeline Company. For eighteen years, through all kinds of weather, the assembly of pipes and valves—the Christmas tree—had stood in that small clearing in the forest. Only now had the rising price of natural gas made the well an economic asset, justifying the construction of a pipeline to tap it.

Halbouty's grin split his face and his dark eyes sparkled as he put down the receiver. "What about that, Doyle?"

McClennen's grin matched Halbouty's. "Nobody can say we haven't got production in Alaska."

"No, by God!" Halbouty said. "After eighteen years we've got it! This wildcatter's got production in Alaska!"

He stood up and slapped his hand on his desk. "And we're going to get some in the Lower Cook Inlet, Doyle! You wait and see!"

Through the years, Halbouty never deserted the position he had explained to Lee J. Cobb in 1969: The Arab oil states dream of world domination and will do their best to achieve it. Even the Egyptian-Israeli peace talks of 1978 did not change his mind. Indeed, his belief was strengthened when emissaries from some Arab oil states offered the Egyptians fifty billion dollars to terminate peace negotiations. It was Halbouty's thought that the oil states wanted a solid Arabic front against Israel, western Europe, and the United States, but would do without Egypt if necessary.

"The time will come," he said over and over, "when the oil states will tell the Egyptians to close ranks or go it alone."

No combination of western powers could offer Egypt financial aid on a scale the oil states could muster, he said, and money, in the end, would probably win the day.

A ray of hope, but not a bright one, was the discovery in Mexico in the late 1970s of vast oil reserves whose potential was estimated at two hundred billion barrels. Mexico, while not a member of OPEC, would sell its oil at OPEC prices, whatever they were, and the United States would have to compete with other nations in purchasing it, he maintained. But the Mexican discoveries were substantiating evidence to him that great quantities of hydrocarbons remained to be found worldwide by those willing to hunt for them.

Late in the year a young reporter visited Halbouty to ask a few questions about his Lower Cook Inlet drilling plans. Halbouty satisfied him in a brief interview. The reporter got up from his chair to leave, and Mary Stewart walked into the office to take some dictation. "You're so busy, Mr. Halbouty," the reporter said as he turned toward the door. "I wonder . . . do you ever take a vacation?"

It was a rhetorical question, but Halbouty sat down behind his desk and said, "Yeah, Mr. Yount made me take one."

The reporter didn't know who Mr. Yount was, but he couldn't properly leave.

"It was in 1933," Halbouty said, "and I caught malaria on

a well we were drilling in Liberty County on Double Gum Bayou."

"That long ago?" the reporter asked as he came back to his seat in front of Halbouty's desk.

Stewart began taking down Halbouty's words in shorthand.

"It was on a farm owned by a man named E. W. Boyt," Halbouty said, looking back into the past (while Stewart marveled at his memory). "He was a friend of Mr. Yount's, and we drilled the well for that reason. There was no geology on it. Mr. Yount just wanted to do a favor for a friend.

"Damn, it was hot down there on Double Gum Bayou," Halbouty said, talking more to himself than for the reporter. "And the mosquitoes, they'd come in like clouds and they'd drive you out of your mind. I was all covered with whelps and rash, and so was everybody else." He glanced at Stewart. "Don't take notes on this, Mary."

She ignored him.

"I remember we began drilling in April," Halbouty said, ignoring the reporter and Stewart, looking over them into the past. "I worked my tail off on that well all that summer. I lived on the job. I slept, when I could, between the boilers and in the toolhouse. I knew there wasn't a damned thing there, and I told Mr. Yount so. He just said, 'Keep 'em drilling, Mike.'

"Then one day Mr. Yount told me what he had in mind. He wanted to drill down past ten thousand feet." Halbouty laughed. "He wanted to be the first to do it. What he didn't know, and I didn't either, of course, was that two wells already had gone past ten thousand, one in California and one in Mexico. Both of them were drilled in 1931."

He leaned forward to peer at the reporter. "Do you understand what I'm saying? I'm telling you about the state of the drilling art and how it had advanced. Listen! In the late twenties, while I was still at A&M, there was a headline, I mean the big headline, the main one, in the Orange, Texas, newspaper, and it said, 'GULF DRILLS TO ONE MILE!'

"You see, being able to drill past five thousand feet was so damned spectacular that it made headlines. And now here was Mr. Yount wanting to drill past ten thousand and two people already had done it. But we didn't know it, like I said, so it was damned exciting."

He laughed and shook his head. "We were trying to do it with a rig that wasn't built for it. Why, that damned rig wasn't built to go to seven thousand feet, much less ten thousand. And everything in the world happened to that well. Everything! And that heaving shale! I put every kind of mud mixture imaginable down that hole and that shale would just gobble it up. Whew, that was a rough son of a bitch.

"Anyway, in late August—I'm sure it was late August—malaria got me. Those damned mosquitoes. I went plumb out of my head and they had to carry me to the hospital, in Beaumont. I was delirious. Out of my head. I don't know how long, how many days, but when I came out of it Mr. Yount was standing by my bed.

"The first thing I said was, 'Did we get to ten thousand?' Mr. Yount shook his head. He said, 'Not quite, Mike. We missed it by a little more than a hundred feet.' He was right. Later on I checked the log and we had gone down to nine thousand, eight hundred and sixty-three feet! And remember, that rig wasn't designed to go deeper than seven thousand."

Halbouty put on his professor's face. "You understand that even if we didn't make ten thousand, it was still a hell of an accomplishment. And it was defeats like that that brought on new technology and new equipment and in a few years drilling to ten thousand feet was commonplace." He grinned. "It would have been great to have reached ten thousand with that pile of junk!"

"What about the vacation?" the reporter asked.

Halbouty's head jerked. "What?"

"The vacation," the reporter said.

"Oh. Well, Mr. Yount said I had to take one when I got

well, and I did. He gave me a company car to travel in and said he would pay for the trip. He asked me how much money I needed. Hell, I didn't know. I said two hundred dollars ought to do it, but he gave me seven hundred . . . no, seven fifty. Told me to get out of Texas and see something."

"Where did you go?" the reporter asked.

"They were holding a world's fair in Chicago, so I went there first. It was terrible, seeing the people on those streets. I knew there was a Depression on, but the streets were full of people out of work. Sad-eyed people, I mean. You don't know anything about it," he said directly to the reporter, "and I hope you never see days like those."

"Did you go to the fair, Mr. Halbouty?"

"Oh, yes. During the day I'd study the things in the exhibits, and at night I'd go to the fun places. I saw Sally Rand." He grinned. "She was the first fan dancer, and it was supposed to be wicked as hell."

The reporter showed no particular interest in Sally Rand and wickedness. "Where else did you go?" he asked.

Halbouty suddenly was abrupt. "I went to New York. I stayed there three days and I didn't like it. I got the hell out of there and drove home. I still had almost three hundred dollars. I turned it in to the company and got my tail back to work." He stood up. "Well, I got to get to work. Thanks for coming by."

"Thank you," the reporter said, and got up. Stewart followed the reporter to see him on his way. In the reception office he touched her arm. "Do you like working for him?" he asked.

She said she did.

"He's kind of weird, isn't he?"

"Kind of," she agreed.

The reporter laughed. "He talked my arm off about that damned well. Seems like he was more interested in that well than he was in the vacation. Right?"

"Right," she said.

Acknowledgment

Of the dozens of persons interviewed in the preparation of this story about Michel T. Halbouty, none was more helpful than the late James A. Clark, the oil writer and historian, who shared his memories and records with me. Clark sold newspapers on the streets of Beaumont with Halbouty when they were boys, and remained closely associated with the wildcatter until Clark's death in 1978.

I am grateful also to Warren Baker, president of the Energy Research & Education Foundation at Rice University. He and his chief assistant, Barbara Schuessler, gave me hours of their time and access to the EREF library, a storehouse of oil and gas information.

A prime source of material was the country's newspapers, from New York City to Los Angeles and from Laredo, Texas, to Anchorage, Alaska. Large and small, it seems that almost every daily has printed a story about Halbouty at some time during his half century in the oil business. Oil editors of the

Houston newspapers—the *Chronicle* and the *Post*—find that Halbouty provides a steady flow of news for their pages. And so do the oil trade periodical editors, particularly those at *The Oil Daily* and the *Oil and Gas Journal.*

Also of great use were the various published reports and hearings of the Multinationals Subcommittee of the Senate Foreign Relations Committee. The energy crisis and embargo of 1973 prompted the committee to investigate the relationship between the importers of foreign oil and the OPEC states. This material is available from the Government Printing Office, Washington, D.C.

Other sources were the *United States, OPEC and Multinational Oil* by Frank R. Wyant (Lexington, Mass.: D. C. Heath and Company); *The Growth of Integrated Oil Companies,* by John G. McLean and Robert W. Haigh (Cambridge, Mass.: Harvard University Graduate School of Business Administration); and various issues of the American Association of Petroleum Geologists *Bulletin.*

The chief source, however, was Michel T. Halbouty himself, who, like most of us, remembers quite vividly that which he wishes to recall and forgets quite easily that which he wishes to ignore.

Index

Ferber, Edna, 118
Fields, oil:
  Ashland, 97
  Cotton Lake, 62
  East Texas, 19, 27, 32, 35,
    40-41, 88, 119, 169-170,
    180, 191
  Escobas, 203
  False River, 227
  Fort Trinidad, 204
  Fostoria, 135-137
  Golden Lane reef, 225
  Hastings, 81
  Justina, 123
  Katy, 201-202
  Lafitte, 79, 81
  Lake End, 98, 100
  Madisonville, 204
  Normanna, 154
  North Lochridge, **123**
  Pine Island, 100-102, 119-120,
    121, 204, 206
  Poesta Creek, 123
  Ryckman, 231
  South Boling, 113-115, 122,
    150
  South Liberty, 102, 104,
    105-108, 111, 114-115,
    120
  Swanson River, 160, 164
  thrust belt fields, 231
  West Beaumont, 58, 64-65, 250
  West Hackberry, 134-135
Ford, Gerald, 219-224
Foreign oil, dependence on,
    184-186, 188-189, 192-193,
    194, 210-212, 219-223, 238,
    255
  Six Day War crisis, 191
  Suez crisis, 176-179
Foster, Robert, 86-87
Freedman, Jake, 107-108
Friendswood, 68-69, 80

Gainer, Charles, 104-105
Galbraith, John Kenneth,
    208-209
Gardner, Herman, 120
Gas, gas fields and wells, 131-134,
    142-150, 157, 214-215
  Kenai field, 163
  Pheasant field, 133-134
  West Fork well, 167
Giant (Ferber), 118
Ginther, Noble and Wilbur, 200
Glen Rose limestone, 204,
    225-226, 227
Gonzaullas, Lone Wolf, 41-43
Grinsfelder, Sam, 163
Gulf Oil Company, 180, 200, 256
Gulf Publishing Company, 59,
    207

Haas, Merrill, 195
Halasko, 159, 254
Halbouty, Fay Renfro Kelly,
    90-92, 93, 95-98, 101,
    102-103, 106, 113, 119-120,
    123, 128, 153, 158, 248-249
Halbouty, Jim, 79, 101
Halbouty, Lesly Carlton, 47,
    58-59, 73, 74, 75, 82, 84,
    85-86
Halbouty, Linda Fay, 120-121,
    128, 247
Halbouty, Michel Thomas:
  his acidizing method, 101-102
  Alaskan venture, 158-167, 205,
    231-232, 253-254
  in the army, 84-92
  awards and medals, 224-225,
    227-228, 232-233, 249-253
  in banking, 170-173, 217
  becomes a millionaire, 114-115,
    122
  beginning successful career on
    his own, 92-98